T0066724

Geophysics: A Very Short Introduction

VERY SHORT INTRODUCTIONS are for anyone wanting a stimulating and accessible way into a new subject. They are written by experts and have been translated into more than 45 different languages.

The series began in 1995, and now covers a wide variety of topics in every discipline. The VSI library currently contains over 550 volumes—a Very Short Introduction to everything from Psychology and Philosophy of Science to American History and Relativity—and continues to grow in every subject area.

Very Short Introductions available now:

Available soon:

For more information visit our website

www.oup.com/vsi/

William Lowrie

GEOPHYSICS

A Very Short Introduction

Great Clarendon Street, Oxford, OX2 6DP,
United Kingdom

Oxford University Press is a department of the University of Oxford.
It furthers the University's objective of excellence in research, scholarship,
and education by publishing worldwide. Oxford is a registered trade mark of
Oxford University Press in the UK and in certain other countries

© William Lowrie 2018

The moral rights of the author have been asserted

First edition published in 2018

Impression: 5

All rights reserved. No part of this publication may be reproduced, stored in
a retrieval system, or transmitted, in any form or by any means, without the
prior permission in writing of Oxford University Press, or as expressly permitted
by law, by licence or under terms agreed with the appropriate reprographics
rights organization. Enquiries concerning reproduction outside the scope of the
above should be sent to the Rights Department, Oxford University Press, at the
address above

You must not circulate this work in any other form
and you must impose this same condition on any acquirer

Published in the United States of America by Oxford University Press
198 Madison Avenue, New York, NY 10016, United States of America

British Library Cataloguing in Publication Data
Data available

Library of Congress Control Number: 2017960669

ISBN 978-0-19-879295-6

Printed and bound by CPI Group (UK) Ltd, Croydon, CR0 4YY

Links to third party websites are provided by Oxford in good faith and
for information only. Oxford disclaims any responsibility for the materials
contained in any third party website referenced in this work.

Contents

Acknowledgements

Alan Green read and criticized the first draft of this work and made many corrections and suggestions. I hope he enjoyed our many lively discussions as much as I did. My wife Marcia drew my attention to sections that were not clear or would be difficult for a non-scientist to understand. I thank Alan and Marcia for their invaluable help. I am grateful to Markku Poutanen of the Finnish Geospatial Research Institute; Christoph Förste of GFZ, German Research Centre for Geosciences, Potsdam; and Dmitry Storchak of the International Seismological Centre, for kindly providing the illustrations of Fennoscandian uplift, the EIGEN6S4 geoid, and global seismicity, respectively. I also thank an anonymous reader for helping me to correct and improve the final text.

List of illustrations

Geophysics

Chapter 1
What is geophysics?

The mining towns in northern Canada are remote from good roads but they are surrounded by lakes. Aircraft, mounted on floats instead of wheels, serve as a practical means of transport in the forested wilderness. After the ice had melted from the lakes in the early summer of 1961, I found myself—a young Scot with a good recent degree in physics but minimum training for the task ahead—disembarking from a float-plane on the shore of a large lake in northern Manitoba, 80 kilometres from the nearest town where the mining company that employed me had its headquarters. My companions were five Cree Indians and a cook, and for the rest of the short northern summer my job would be to use geophysical equipment to search for potential sources of nickel in the geological structures buried beneath the forest floor. This was my initiation to the world of geophysical exploration. It convinced me to change from physics to a career as a geophysicist. In addition to teaching and laboratory research it involved annual fieldwork that took me into less developed areas of the world where interesting geological problems could be addressed.

Geophysics is a field of earth sciences that uses the methods of physics to investigate the physical properties of the Earth and the processes that have determined and continue to govern its evolution. Geophysical investigations cover a wide range of research fields, extending from surface changes that can be observed from Earth-orbiting satellites to unseen behaviour in the Earth's deep interior. The properties of the Earth are complex, such that sophisticated methods are needed to study its natural processes. In contrast to a physics experiment, which can be conducted under carefully controlled laboratory conditions, a geophysical investigation is carried out under circumstances that are set by nature and cannot be completely controlled. This is one reason why earthquake prediction is not yet possible, despite enormous endeavour by seismologists. An additional complicating factor is time: simulations of geological processes that have taken place over thousands of years have to be computed in a short experimental time.

The timescale of processes occurring in the Earth has a very broad range. It includes rapid events like the violent shaking of an earthquake, which may last fractions of a second to several minutes, depending on the severity of the earthquake. However, most geological processes take place very slowly, on very long timescales. For example, slow changes in the Earth's magnetic field—such as polarity reversals—happen over thousands of years, and the motions of tectonic plates take place over tens of millions of years. Yet long timescale processes leave measurable traces in the rock record that can be analysed and understood using geophysical methods.

Although it is often regarded as an offshoot of physics, geophysics handles topics that interested philosophers long before modern physics evolved as a separate subject in the 19th century. Until then, physics and chemistry belonged to 'natural philosophy', the study of nature by logical and quantitative methods. The scientific revolution started in the 16th century when Nicolaus Copernicus formulated the heliocentric model of the solar system, in which the planets move around the Sun—rather than the Sun and planets moving around the Earth, which had previously been the orthodox belief. The revolution, which extended into the 17th century, was based on the derivation of fundamental laws from observations of the natural world by scientists such as Johannes Kepler, Galileo Galilei, William Gilbert, and Isaac Newton. Their respective studies of planetary behaviour, astronomy, geomagnetism, and gravity were among the pioneering contributions to knowledge of the physical world. The principal disciplines of geophysics still include planetary gravitational and magnetic fields. The development of seismology in the 20th century enabled investigations of the Earth's internal structure and composition. Coupled with measurements of the heat flowing out of different regions of the planet's surface, this knowledge has led to an improved understanding of the Earth's internal dynamics.

In the 1960s, data acquired from global seismology and marine geophysical investigations led to the theory of plate tectonics, which provides an explanation for the structure and evolution of the Earth, including displacements of the continents and the origin of tectonically active zones. This caused a revolution in the way we understand the mobility of the Earth's surface. It occurred in parallel with advances in digital electronics that created a massive increase in computing power, enabling sophisticated data processing and computer modelling of geological processes. Other technological improvements have made it possible to acquire, store, and access huge amounts of geophysical data, such as satellite observations of topography, gravity, and the magnetic field.

The ability to measure physical properties from space has led to substantial advances in geodesy—the study of the Earth's shape and gravitational field—with important consequences for interpreting gravity. Gravity depends on the Earth's shape and thus varies with position and altitude, both of which must be determined precisely. Historically, geodesy required painstaking measurements to chart the Earth's shape and to measure the distances between places and their heights above sea level. Much of the Earth—for example, the 71 per cent covered by oceans—was inaccessible for such measurements. Space geodesy overcomes these obstacles by providing high-quality data for the entire globe in a short time. The most familiar type of geodetic data from space is provided by the Global Positioning System (GPS) used in millions of portable navigation instruments. The scientific versions of GPS devices, linked in networks and recording continuously, provide location data that enable the measurement of tiny millimetre-scale displacements of the Earth's surface. These include the ongoing vertical uplift of regions that were depressed by ice sheets during the latest ice age, as well as the horizontal motions at tectonic plate boundaries. These motions occur at rates in the range of a few millimetres to centimetres per year.

The best-known field of geophysics is seismology. Earthquakes are among the greatest hazards to mankind, but the study of how seismic waves generated by these events travel through the Earth has revealed the concentric shell structure of its core, mantle, and crust. Seismometers—the instruments that record an earthquake's tremors—were invented in the 19th century. Originally they could record only limited ranges of the wide spectrum of frequencies of these ground vibrations. The cold war necessitated the development of seismometers that could be used to supervise nuclear test-ban treaties. They had to be able to distinguish between a small nuclear test and a small earthquake. Progressively, seismometers were developed that could record the entire spectrum of frequencies in an earthquake. The deployment of these 'broadband' seismometers improved the precision with which earthquakes can be located, observed, and measured. The development of powerful computers and advanced techniques of data processing have led to an improved understanding of earthquakes and to the growth of a global network of seismic stations for monitoring them. The modern generation of seismometers is so sensitive that even the noisy background signal on seismic records can now be analysed in terms of crustal and upper mantle features.

Commercial companies searching for petroleum and mineral resources have adapted and refined many geophysical methods for their own specific needs. The feedback from their efforts has greatly enriched geophysics by improving instrumentation, data processing, and analytical methods. Some geophysical techniques developed for industrial studies have been adapted to document and solve environmental problems, which occur primarily in the shallow layers of the subsurface. Applied and environmental geophysics are not treated in any detail here, but they are important topics in their own right and merit separate treatment to do them justice. This book presents a general overview of the principal methods of geophysics that have contributed to our understanding of Planet Earth and how it works.

Chapter 2
Planet Earth

Physical laws

A photograph from space of the night side of the Earth shows vividly the effects of urbanization. Brilliant illuminated patches stand out against a dark background, which once would have typified the entire globe. On the Earth, far from this anthropogenic light, the sky on a dark night is a cause for wonder; it probably fascinated our ancestors from time immemorial. Thousands of years ago Chinese astronomers defined the year and month from the repeated motions of the Sun and Moon, respectively. The stars form an apparently steady firmament, but from early times astronomers noted that some stars seemed to move against this background. The word 'planet' derives from the name used by the ancient Greeks for these 'wandering stars'. The planets Mercury, Venus, Mars, Jupiter, and Saturn are visible to the naked eye; Uranus is very faintly visible to the naked eye, but in 1781 it became the first planet to be confirmed by telescope. Since ancient times astronomers have noted and documented the motions of the planets. They recognized that the motions are systematic and follow fundamental rules.

Two important laws of physics determine the behaviour of the Earth as a planet and the relationship between the Sun and its planets. First, in an isolated system, in which there is no addition or loss of energy with respect to external sources, the total energy of the system is constant. This is known as the law of conservation of energy. It means that energy is neither created nor destroyed, although it can be transformed from one form to another. For example, the combustion of coal produces heat (a chemical transformation), which can convert water to steam (a change of state); this can drive a turbine, thus converting thermal energy to the energy of motion (called kinetic energy), and eventually generating electrical power.

The second law defines the conservation of angular momentum. The momentum of an object moving in a straight line is defined as the product of its mass and velocity. For a rotating object, the manner in which its mass is distributed about the axis of rotation, together with the rate of rotation, define the object's angular momentum. For a tiny point mass the angular momentum is the product of its linear momentum and its distance to the axis about which it is rotating. In the case of an extended object, the angular momentum is the sum of such quantities for each particle of the object. The angular momentum of an isolated system is constant. However, the speed of rotation can still change; for example, if the distance of the object's mass to the rotation axis changes. A familiar example is the pirouette in figure skating: as the skater's arms are drawn inwards, reducing their distance to the rotation axis, the speed of rotation increases so as to maintain constant angular momentum.

The solar system

According to the 'Big Bang' model, the universe is believed to have originated 13.8 billion years ago in a state characterized by high temperature and highly concentrated energy. The energy expanded rapidly into the surrounding space, effectively decreasing its density and temperature. Many individual processes involving subatomic particles took place successively in the first seconds. Only a few minutes after the Big Bang, nuclei of hydrogen and helium were formed; these elements constitute, respectively, about 73 per cent and 25 per cent of the mass of the present known universe; the remaining 2 per cent are the elements heavier than helium. This theory provides a basis for understanding how the solar system, and in particular the Earth, formed.

The Sun is a star that originated together with the planets about 4.5 billion years ago in a vast cloud of molecular hydrogen and interstellar dust, ranging from a few molecules to sub-millimetre size. Gravitational attraction drew the particles towards their common centre of mass, eventually forming a star, the Sun. It is unlikely that the particle motion was purely radial, so there would have been a net component of rotation. As the cloud collapsed inwards, distances from the rotation axis decreased and the rotation speed of the cloud increased so as to conserve angular momentum. The rotating mixture of gas and dust flattened to form a disc with the newly formed Sun at its axis of rotation. Collisions between particles were inevitable in the swirling mass of material. They resulted in the coalescence of particles to form larger objects. This eventually led to the accumulation of kilometre-sized planetesimals, which in turn accreted to form even larger objects, called protoplanets. These were several hundred kilometres in diameter, similar in size to the moons of Mars. Collisions between the protoplanets, in cooperation with mutual gravitational attraction, eventually led to the formation of the planets. It has been suggested that early in its history the Earth collided with a hypothetical planet similar in size to Mars, creating a ring of debris that eventually coalesced to form the Earth's only natural satellite, the Moon.

The Earth's own gravity caused the planet to compact, which released heat. This was augmented by heat from radioactive decay, so that eventually the internal temperature reached the melting point of iron. Gravity caused the heavier elements (predominantly iron and nickel) to sink to form a dense core, while lighter elements moved upwards to form a silicate mantle around it. A chemically different thin crust formed later on the surface of the mantle, and may have been renewed many times. In this way the Earth acquired a layered structure similar to that of a soft-boiled egg: a thin hard shell surrounds a firm mantle, within which there is a liquid core. A layered internal structure is found in other planets and their moons, but these are less well determined than the Earth's.

The distances of the planets from the Sun (Table 1) increase with remarkable regularity. If the asteroid belt between Mars and

Table 1. Dimensions of planetary orbits

Planet	Mean radius (AU)	Eccentricity	Sidereal period (yr)
Terrestrial planets and Moon			
Mercury	0.383	0.206	0.241
Venus	0.723	0.0067	0.615
Earth	1.000	0.0167	1.000
Moon	0.00257	0.0549	0.0748
Mars	1.52	0.0935	1.88
Great planets			
Jupiter	5.20	0.0489	11.9
Saturn	9.57	0.0565	29.5
Uranus	19.2	0.0457	84.0
Neptune	30.0	0.0113	165
Dwarf planet			
Pluto	38.9	0.249	248

An astronomical unit (AU) equals 149.6 million km, approximately the mean radius of the Earth's orbit. Orbital periods are in Earth-years.

Source for more detailed information: <http://nssdc.gsfc.nasa.gov/planetary/>.

Jupiter is included, each orbit has roughly double the radius of its nearest neighbour. Although it is unlikely that the regularity is by chance, no satisfactory physical explanation has yet been able to account for it. A further mystery of the solar system is the unequal distributions of mass and angular momentum. Although the Sun contains more than 99 per cent of the mass of the solar system, planetary motions represent more than 99 per cent of the angular momentum. The combined angular momentum of all the planets defines a plane, called the invariable plane of the solar system, which may at one time have been the protoplanetary disc—that is, the rotating disc of gas and dust from which the solar system evolved.

Eight bodies orbiting the Sun are currently recognized as planets. They form two categories (Figure 1) based on their size, composition, and distance from the Sun. The distances are

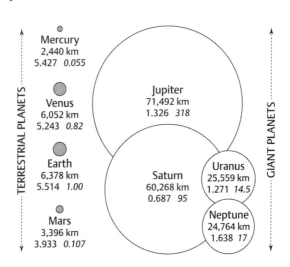

TERRESTRIAL PLANETS

GIANT PLANETS

Mercury
2,440 km
5.427 0.055

Venus
6,052 km
5.243 0.82

Earth
6,378 km
5.514 1.00

Mars
3,396 km
3.933 0.107

Jupiter
71,492 km
1.326 318

Saturn
60,268 km
0.687 95

Uranus
25,559 km
1.271 14.5

Neptune
24,764 km
1.638 17

1. **Relative dimensions of the planets. The first number beneath the planet's name is its equatorial radius, the second number is its mean density relative to water, the third number (in italics) is its mass relative to the Earth.**

conveniently measured in astronomical units (AU)—this is roughly the mean radius of the Earth's orbit around the Sun. The inner, or terrestrial, planets (sequentially Mercury, Venus, the Earth, and Mars) are closest to the Sun and are composed of rock and metal; they are relatively small and are surrounded by few moons and no rings. The outer, or giant, planets (in order of increasing distance: Jupiter, Saturn, Uranus, and Neptune) formed further from the Sun, in colder parts of space. Jupiter and Saturn are gas giants. More than 90 per cent of their masses consists of hydrogen and helium; their atmospheres also contain ammonia and water and they have metallic hydrogen cores. Uranus and Neptune are ice giants. Only about 20 per cent of their masses consists of hydrogen and helium; beneath their gaseous atmospheres they consist of ices of water, methane, and ammonia. The giant planets have many moons and are also surrounded by systems of rings of dust. Between Mars and Jupiter lies the asteroid belt, which consists of numerous objects with terrestrial compositions. The smallest are particles of dust; the four largest are several hundred kilometres in diameter and are referred to as minor planets. The largest of these, Ceres, is 950 km in diameter and is termed a dwarf planet. All the planets have slightly elliptical orbits that lie within a few degrees of the Earth's orbital plane, which is called the ecliptic. This differs by only about 1 degree from the invariable plane and is used as reference plane for the solar system.

The Sun is orbited by countless other bodies. They lie beyond Neptune and are known collectively as Trans-Neptunian objects (TNO). Thousands of these are icy planetesimals and minor planets forming a disc-shaped region called the Kuiper belt, which lies close to the ecliptic plane at 30 AU to 50 AU from the Sun. The best known of the TNO is Pluto. Long regarded as the furthest planet from the Sun, it has been demoted to the status of dwarf planet because of its small size; it is smaller than the Moon and has only 4 per cent the mass of Mercury, the smallest recognized planet.

Kepler's laws of planetary motion

Gravitation is one of the fundamental forces in physics. It is the force of attraction between two objects that arises solely from their masses and varies inversely with the square of their separation. This is the law of universal gravitation, published in 1687 by Isaac Newton. The space in which the attraction of a mass is felt is called its gravitational field.

In the late 16th century, the astronomer Tycho Brahe made accurate observations of the positions of the planets using an astrolabe, an ancient astronomic instrument that was in use since the era of classical Greece. Although Brahe's measurements were made before the telescope was invented, they were so precise that, in 1609 and 1619, Johannes Kepler could combine them into three laws of planetary motion (Figure 2). These are: (1) each planetary

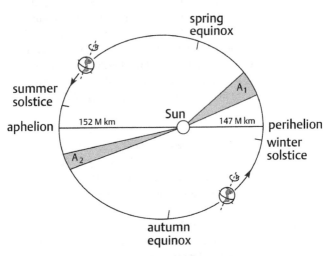

2. Schematic drawing of the Earth's elliptical orbit, showing the positions of aphelion and perihelion, the equinoxes and solstices for the northern hemisphere. Distance unit is 1 million km (M km).

orbit is an ellipse with the Sun at one of its focal points, (2) the radius connecting the planet to the Sun sweeps out equal areas in equal time intervals, and (3) the square of the period of the motion is proportional to the cube of the semi-major axis of the ellipse. In contrast to a circle, which has a constant radius, the shape of an ellipse is defined by its shortest (minor) axis and its longest (major) axis. The deviation of an ellipse from a circle is known as its eccentricity; by definition it is 0 for a circle. As the eccentricity increases, the ellipse becomes progressively more elongate. Except for the orbit of Mercury (which has an eccentricity of 0.2) the planets have nearly circular orbits.

The motion of a planet about the Sun can be regarded for simplicity as a closed system, free of external influences. In fact, the other planets do influence each other's motion, but to a lesser degree than the Sun does. Each planet orbits the Sun under its gravitational attraction, which acts along a radius of its orbit. The radial force results in a constant angular momentum of the planet about the Sun. As a result, the orbit of the planet is a plane that passes through the Sun. The planet's orbital motion in this plane is subject to the conservation of energy. Two types of energy are relevant to this motion: the energy of the gravitational attraction that binds the planet to the Sun, called its potential energy, and the energy associated with the speed of its motion, called its kinetic energy. If the planet is moving so fast that the kinetic energy is greater than the potential energy, it escapes from the solar system on a curved trajectory known as a hyperbola. If the potential energy is dominant, such that the planet cannot escape from its orbit, and if the force binding it to the Sun varies as the inverse square of distance (as with gravitation), then the orbit takes the shape of an ellipse with the Sun at one of its focal points. This is Kepler's first law.

The Earth has a slightly elliptical orbit, which means that it is about 147 million km from the Sun at its closest point—known as perihelion—while the distance is 152 million km at its furthest point—called aphelion (Figure 2). Perihelion is passed each year around 3 January, and aphelion around 3 July. The line joining the extreme points is the major axis of the ellipse, and is called the line of apsides.

Kepler's second law follows from the conservation of a planet's angular momentum about the Sun. It results in equal areas (e.g. A_1 and A_2 in Figure 2) being swept by the radius vector in the same time. Consequently, a planet's speed around its orbit is variable: at perihelion, it is moving faster than at aphelion. Kepler's third law results from combining the period of the variable planetary motion with the equation of an ellipse (the first law). In 1687 Isaac Newton showed that Kepler's first and third laws confirmed his inverse square law of universal gravitation.

The Earth spins about an axis that is inclined to the pole of the ecliptic plane; the axial tilt is called the obliquity of the ecliptic (Figure 3A). The angle at present measures 23.44 degrees but varies slowly with a period of 41,000 yr as a result of interactions with other planets. The obliquity is responsible for the annual seasons. The motion of the Earth around its orbit changes the attitude of each hemisphere to the Sun, affecting the length of the day. Shortly before perihelion, the obliquity causes the northern hemisphere to be tilted away from the Sun, with the shortest day at the winter solstice on 21 December. Six months later the northern hemisphere is tilted towards the Sun, and the day is longest at the summer solstice on 21 June. The line joining these positions is called the line of solstices. Twice each year the rotation axis is normal to the radius to the Sun, so that around 21 March and 23 September the day and night are equally long; these positions are called the spring and autumn equinoxes and the line joining them is called the line of equinoxes. The axial tilt causes the seasons to differ by six months between the northern and southern hemispheres. Summer in the southern hemisphere occurs near perihelion, so it should be warmer than summer in the northern hemisphere; correspondingly, southern hemisphere winters fall near aphelion and should be colder. However, the southern hemisphere is dominated by oceans and the northern hemisphere by land. The oceans heat up and cool off more slowly than land surfaces, and as a result climates are milder in the southern hemisphere.

The Chandler wobble

The Earth is not a rigid body but reacts elastically in response to deforming forces. Its ideal shape would be a sphere, but the centrifugal forces arising from its rotation cause the shape to flatten slightly about the spin axis, so that the equatorial diameter is longer than the polar diameter. The slightly flattened sphere is called an oblate spheroid. The flattening is not large, only about 1 part in 300, but it affects how the Earth rotates.

For example, if the axis of the spinning Earth is displaced, then allowed to spin freely, it exhibits a wobbling motion about its mean location, in the same way as a spinning top wobbles, if nudged. This motion results from the Earth's unequal internal mass distribution and is called a free nutation, or 'nodding motion'. The mean direction of the Earth's rotation axis is constant but at any instant it may be displaced slightly from its mean direction. The instantaneous axis moves around the mean rotational pole with an average displacement of several metres. The free nutation was explained mathematically in 1765 by Leonhard Euler, a Swiss mathematician, but was only detected in 1891 by Seth Chandler, an American astronomer, after whom it is called the Chandler wobble. Euler predicted a period of about ten months for the motion, but the observed period is fourteen months. The 40 per cent increase in period is due to the elastic yielding of the Earth, which allows it to deform to accommodate the displacement of the instantaneous rotation axis; Euler's model had assumed a rigid planet. The mechanism, or 'nudge', that excites the Chandler wobble has not been identified conclusively. Proposed sources have included very large earthquakes, atmospheric fluctuations, and ocean-bottom pressure changes related to oceanic circulation.

The development of Very Long Baseline Interferometry (VLBI) has made it possible to observe and measure the Chandler wobble precisely. This technique, used in radio astronomy, has been adapted for geodetic purposes as follows. Radio sources—e.g. quasars—that lie outside the Milky Way galaxy provide a very stable coordinate system against which the motion of the planet can be measured. The extraterrestrial radio signals are detected by radio telescopes combined into large arrays at different places on the Earth. The time differences between the arrival times of repeated signals at separate arrays are analysed to obtain the orientation of the Earth and its rotational rate with exceptional accuracy. The VLBI data trace the Chandler wobble

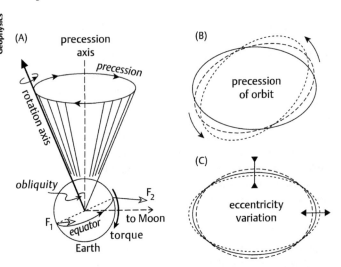

3. (A) Precession of the Earth's rotation axis induced by the Moon's torque on the equatorial bulge. (B) Precession of the major axis of the Earth's orbit relative to a stellar reference frame. (C) Variation of the eccentricity of the orbit. The diagrams are not to scale.

with a precision of a few centimetres, and the rotational period is determined to better than 0.1 msec. This enables precise observation of changes in the length of the day and the identification of contributing factors, such as the rotational braking caused by marine tides, changes in the angular momentum of the atmosphere, and the effects of bodily tides in the solid Earth.

Effects of the Moon and Jupiter on the Earth's rotation

The Moon's gravitational attraction disturbs the orientation of the Earth's rotation axis. Due to the obliquity of the Earth's axis the equatorial bulge projects below the plane of the Moon's orbit on one side of the Earth and above it on the opposite side. The Moon's gravitational attraction on the distant side is weaker than on the closer side, because of the inverse square law. This results in a torque, or turning force, which attempts to bring the rotation axis upright, that is, normal to the ecliptic (Figure 3A). However, when a torque acts on a spinning object, it causes the rotational axis to move while maintaining the angle of tilt (the obliquity in this case). As a result, the rotation axis moves around the surface of a cone that has the pole to the ecliptic as its axis. This motion is called precession. Viewed from above the ecliptic, the Earth's rotation is anticlockwise, while the direction of precession is clockwise. Because it takes place in the opposite sense to the Earth's rotation, the precession is said to be *retrograde*. The gravitational attraction of the Sun on the equatorial bulge also causes retrograde precession. Although the Sun is vastly more massive than the Moon, it is much further from the Earth and as a result its effect on the precession is about half that of the Moon. The combined precession is known as the precession of the equinoxes, and its period is approximately 25,800 yr.

The Earth exerts a reciprocal torque on the Moon, which results in a precession of the lunar orbit around the Earth with a period of 18.6 yr. This then modulates the amplitude of the lunisolar precession of the Earth's axis by superposing a small forced nutation. This changes the angle of obliquity by a tiny fluctuating amount—up to 9 seconds of arc—with a period of 18.6 yr equivalent to that of the lunar orbital precession. However, this effect is insignificant compared to the main lunisolar precession.

The gravitational attractions of the other planets—especially Jupiter, whose mass is 2.5 times the combined mass of all the other planets—influence the Earth's long-term orbital rotations in a complex fashion. The planets move with different periods around their differently shaped and sized orbits. Their gravitational attractions impose fluctuations on the Earth's orbit at many frequencies, a few of which are more significant than the

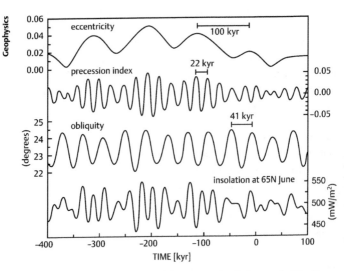

4. Cyclical variations in eccentricity, precession index, and obliquity, with the resulting fluctuations in insolation calculated for the past 400,000 years and the next 100,000 years. Time advances from left to right on the horizontal axis.

rest. One important effect is on the obliquity: the amplitude of the axial tilt is forced to change rhythmically between a maximum of 24.5 degrees and a minimum of 22.1 degrees with a period of 41,000 yr. Another gravitational interaction with the other planets causes the orientation of the elliptical orbit to change with respect to the stars (Figure 3B). The line of apsides—the major axis of the ellipse—precesses around the pole to the ecliptic in a *prograde* sense (i.e. in the same sense as the Earth's rotation) with a period of 100,000 yr. This is known as planetary precession. Additionally, the shape of the orbit changes with time (Figure 3C), so that the eccentricity varies cyclically between 0.005 (almost circular) and a maximum of 0.058; currently it is 0.0167 (Table 1). The dominant period of the eccentricity fluctuation is 405,000 yr, on which a further fluctuation of around 100,000 yr is superposed, which is close to the period of the planetary precession.

Milankovitch cycles of climatic variation

The amount of solar energy received by a unit area of the Earth's surface is called the insolation. This can be calculated for the top of the atmosphere from the known solar irradiation. If the atmosphere were transparent, the surface insolation would only be affected by the separation of the Earth and the Sun and the orientation of the surface to the solar energy. These factors change during the year as the Earth moves around its orbit; the distance from the Sun varies and the axis tilts alternately towards the Sun, then away from it, causing the changing seasons.

The long-term fluctuations in the Earth's rotation and orbital parameters influence the insolation (Figure 4) and this causes changes in climate. When the obliquity is smallest, the axis is more upright with respect to the ecliptic than at present. The seasonal differences are then smaller and vary less between polar and equatorial regions. Conversely, a large axial tilt causes an extreme difference between summer and winter at all latitudes. The insolation at any point on the Earth thus changes with the obliquity cycle. Precession of the axis also changes the insolation. At present the north pole points away from the Sun at perihelion; one half of a precessional cycle later it will point away from the Sun at aphelion. This results in a change of insolation and an effect on climate with a period equal to that of the precession. The orbital eccentricity cycle changes the Earth–Sun distances at perihelion and aphelion, with corresponding changes in insolation. When the orbit is closest to being circular, the perihelion–aphelion difference in insolation is smallest, but when the orbit is more elongate this difference increases. In this way the changes in eccentricity cause long-term variations in climate. The periodic climatic changes due to orbital variations are called Milankovitch cycles, after the Serbian astronomer Milutin Milankovitch, who studied them systematically in the 1920s and 1930s.

The evidence for cyclical climatic variations is found in geological sedimentary records and in long cores drilled into the ice on glaciers and in polar regions. The weathering and erosion of rocks produce mineral particles that are transported and deposited in lakes and seas to form sediments. Other sources of sediment are the accumulation of shells of marine organisms and the chemical precipitation of minerals. These sedimentary processes are affected by rainfall and temperature and thus react to changes in climate. Sedimentation takes place slowly over thousands of years, during which the Milankovitch cycles are recorded in the physical and chemical properties of the sediments. Analyses of marine sedimentary sequences deposited in the deep oceans over millions of years have revealed cyclical variations in a number of physical properties. Examples are bedding thickness, sediment colour, isotopic ratios, and magnetic susceptibility.

When ice accumulates on a glacier or in polar regions, it absorbs oxygen from the atmosphere. Important climate records have been obtained from studying isotopes of the oxygen in polar ice cores. Isotopes are separate forms of an element that differ only in the number of neutrons they contain. The two most common isotopes of oxygen contain 16 and 18 neutrons, respectively; their ratio in water depends on the temperature. The records of oxygen isotope ratios in long ice cores display Milankovitch cycles and are important evidence for the climatic changes, generally referred to as orbital forcing, which are brought about by the long-term variations in the Earth's orbit and axial tilt.

Apart from its climatic significance, the pattern of Milankovitch cycles in a sediment allows it to be dated more completely. The age of a sedimentary sequence is often known at only a small number of widely spaced depths, for example at levels that have been dated radiometrically. Between these dated horizons there may be distinctive undated features such as faunal extinctions or magnetic polarity reversals, which occur at irregular intervals. The presence of cyclical variations in a physical property of a sediment makes it possible to estimate the ages of palaeontological or magnetic events between the dated levels in a sedimentary record.

Chapter 3
Seismology and the Earth's internal structure

Elastic deformation

Seismology is the most powerful geophysical tool for understanding the structure of the Earth. It is concerned with how the Earth vibrates. In the same way that the strings of a guitar vibrate back and forth in a periodic motion when plucked, the solid material in the Earth reacts to a sudden jolt by vibrating. This is particularly evident when an earthquake strikes, but it can also happen as reaction to a local shock. Physically, seismic behaviour depends on the relationship between stress and strain in the Earth, so if we want to understand seismology, it is worthwhile examining these properties.

Stress is defined as the force acting on a unit area. The fractional deformation it causes is called strain. The stress–strain relationship describes the mechanical behaviour of a material. When subjected to a low stress, materials deform in an elastic manner so that stress and strain are proportional to each other and the material returns to its original unstrained condition when the stress is removed. Seismic waves usually propagate under conditions of low stress. If the stress is increased progressively, a material eventually reaches its elastic limit, beyond which it cannot return to its unstrained state. Further stress causes disproportionately large strain and permanent deformation. Eventually the stress causes the material to reach its breaking point, at which it ruptures. The relationship between stress and strain is an important aspect of seismology. Two types of elastic deformation—compressional and shear—are important in determining how seismic waves propagate in the Earth.

Imagine a small block that is subject to a deforming stress perpendicular to one face of the block; this is called a normal stress. The block shortens in the direction it is squeezed, but it expands slightly in the perpendicular direction; when stretched, the opposite changes of shape occur. These reversible elastic changes depend on how the material responds to compression or tension. This property is described by a physical parameter called the bulk modulus. In a shear deformation, the stress acts parallel to the surface of the block, so that one edge moves parallel to the opposite edge, changing the shape but not the volume of the block. This elastic property is described by a parameter called the shear modulus.

An earthquake causes normal and shear strains that result in four types of seismic wave. Each type of wave is described by two quantities: its wavelength and frequency. The wavelength is the distance between successive peaks of a vibration, and the frequency is the number of vibrations per second. Their product is the speed of the wave. Two types of wave travel through the body of the Earth and two types spread out at and near its surface. They are characterized by different kinds of ground motion and their speeds depend in different ways on the elastic properties and density of the rocks they travel in.

Seismic body waves

Seismic waves are sensed by a device called a seismometer; the recording is called a seismogram, and the combined instrument forms a seismograph. One common type of seismometer consists

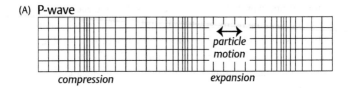

(A) P-wave

compression   particle motion   expansion

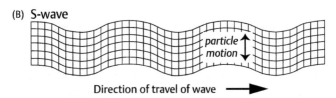

(B) S-wave

particle motion

Direction of travel of wave ➡

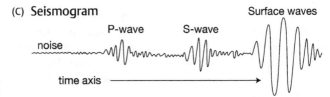

(C) Seismogram

Surface waves

P-wave   S-wave

noise

time axis ➡

5. (A) The energy of a seismic P-wave propagates by a succession of compressions and expansions in the direction of wave propagation. (B) In an S-wave the particle motions are perpendicular to the direction of propagation. (C) Hypothetical seismogram: the P-waves travel fastest and arrive at the seismometer before the S-waves and slower surface waves.

of a heavy magnet within a coil that is attached to a casing connected to the ground. The ground, casing, and coil all move in an earthquake, but the inertia of the heavy magnet restricts its movement, so that relative motion takes place between the magnet and the coil. This induces an electrical current in the coil, which is amplified electronically and recorded. The principle of a modern broadband seismometer is similar: a small mass is held motionless relative to the frame of the instrument by an electronic feedback circuit, which measures the force required to compensate the relative motion caused by ground shaking.

Early seismographs recorded the shaking as a wiggly line traced on paper, but in modern instruments the recording is made digitally and covers a wide range of vibrational frequencies. The seismometer was invented more than a century ago and has been improved continuously, with the result that modern instruments are very sensitive. The device was originally designed to register the ground's displacement, but modern instruments react primarily to the ground's velocity during an earthquake's shaking. They can also be designed to record the ground's acceleration.

A seismic *P-wave* (also called a primary, compressional, or longitudinal wave) consists of a series of compressions and expansions caused by particles in the ground moving back and forward parallel to the direction in which the wave travels (Figure 5A). The speed of a P-wave depends on the bulk modulus, shear modulus, and density of the medium, and is about 6–7 km/s in the Earth's crust (for comparison: the speed of sound in air is 0.33 km/s). It is the fastest seismic wave and can pass through fluids, although with reduced speed. When it reaches the Earth's surface, a P-wave usually causes nearly vertical motion, which is recorded by instruments and may be felt by people but usually does not result in severe damage.

A seismic *S-wave* (i.e. secondary or shear wave) arises from shear deformation (Figure 5B). It travels by means of particle vibrations perpendicular to the direction of travel; for that reason it is also known as a transverse wave. The shear wave vibrations are further divided into components in the horizontal and vertical planes, labelled the SH- and SV-waves, respectively. Shear wave speed depends only on the shear modulus and the density; no volume changes are involved, so, in contrast to the P-wave, the speed does not depend on the bulk modulus. Thus, an S-wave is slower than a P-wave, propagating about 58 per cent as fast, with crustal velocities around 3.5–4 km/s. Moreover, shear waves can only travel in a material that supports shear strain. This is the case for a solid object, in which the molecules have regular locations and intermolecular forces hold the object together. By contrast, a liquid (or gas) is made up of independent molecules that are not bonded to each other, and thus a fluid has no shear strength. For this reason S-waves cannot travel through a fluid. This has important consequences for understanding the internal structure of the Earth. S-waves have components in both the horizontal and vertical planes, so when they reach the Earth's surface they shake structures from side to side as well as up and down. They can have larger amplitudes than P-waves. Buildings are better able to resist up-and-down motion than side-to-side shaking, and as a result SH-waves can cause serious damage to structures.

Seismic surface waves and free oscillations

Surface waves spread out along the Earth's surface around a point—called the epicentre—located vertically above the earthquake's source, in the same way that ripples from a splash spread across a pool. Very deep earthquakes usually do not produce surface waves, but the surface waves caused by shallow earthquakes are very destructive. In contrast to seismic body waves, which can spread out in three dimensions through the Earth's interior, the energy in a seismic surface wave is guided by the free surface. It is only able to spread out in two dimensions and is more concentrated. Consequently, surface waves have the largest amplitudes on the seismogram of a shallow earthquake (Figure 5C) and are responsible for the strongest ground motions and greatest damage.

There are two types of surface wave. A Rayleigh wave combines the longitudinal vibration of a P-wave with the vertical component of vibration of an S-wave, causing the particles of the surface to move around an elliptical path in the vertical plane. If one visualizes a Rayleigh wave moving from left to right, the ground particles move around their ellipses in the anticlockwise direction; this is akin to the motion of particles in a water wave, except that the latter move in a clockwise sense relative to the direction of the wave. They cause a rolling motion of the surface, which may result in destructive ground-shaking. The Rayleigh waves travel at a speed that is about 92 per cent of the S-wave speed. A Love wave arises when the horizontal components of shear waves are trapped in the boundary layer between a free surface and a lower interface. This can produce strong horizontal shaking, with damaging effects on the foundations of structures. The speed of a Love wave is intermediate between that of S-waves in the boundary layer and the deeper interior.

As a result of their different paths and speeds of propagation, the four types of seismic wave take different lengths of time to travel between an earthquake and the seismometer. The first arrival on a seismogram is the P-wave, followed next by S-waves, then the surface waves (Figure 5C). The seismogram is, however, much more complicated than this, because P- and S-waves bounce around in the interior of the Earth, they are bent and reflected at various interfaces, and therefore many arrivals are superposed on a seismogram.

The amplitude of a seismic wave decreases with increasing distance from the earthquake source due to three factors: anelastic attenuation, scattering, and geometrical spreading. Anelastic attenuation is the loss of amplitude of a wave due to absorption of its energy by non-elastic processes as it passes through the Earth. Imperfections in the mineral grains that make up the Earth's interior absorb energy; their effect is complex and is described summarily as internal friction. Scattering occurs when a wave suffers reflection and refraction at irregularities or changes in the material properties of the medium it is passing through. Geometric spreading refers to the distribution of a wave's energy over an increasing area as it spreads. Seismic body-wave energy spreads out on a spherical surface around a deep earthquake. At distance r from the source, the area of a spherical wave surface is proportional to r^2, so the energy attenuates as $1/r^2$. In contrast, the disturbance of a surface wave spreads out as a circular ring around the epicentre, with circumference proportional to r, so its energy only attenuates as $1/r$. Thus, the amplitudes of surface waves decrease more slowly with distance from an earthquake than do the amplitudes of body waves. The energy of surface waves from a very large earthquake can travel around the world several times before it dissipates.

The ground motion in a surface wave is not actually restricted to the surface but penetrates some distance into the Earth, decreasing in amplitude with increasing depth. The 'penetration depth' of a surface wave component is usually taken to be the depth where its amplitude has decreased to about a third of its initial value; it is roughly proportional to the wavelength. A surface wave consists of components with different frequencies, some of which correspond to long wavelengths. Seismic velocities generally increase with depth, so the wave components with long wavelengths penetrate deeper and travel faster than the short wavelength components. Consequently, the shape of a surface wave changes with distance from the source. This phenomenon is called dispersion. Seismologists analyse the dispersion of surface waves to obtain important information about the physical properties and structure of the outer layers of the Earth.

Very large earthquakes can jolt the Earth so severely that the entire globe oscillates, analogous to the ringing of a bell but with periods of several tens of minutes. The frequencies of natural oscillation—also called the normal modes of vibration—are determined by the Earth's elastic properties and internal structure. They occur in three categories according to the type of motion (Figure 6). Radial oscillations consist of displacements along radial directions; the entire Earth alternately swells and contracts, as if it were breathing. Spherical oscillations cause the Earth to bulge outwards at the equator while contracting along the rotation axis, followed a half period later by the opposite displacements at these locations. Toroidal oscillations involve a shear-like behaviour of the globe in which the upper hemisphere

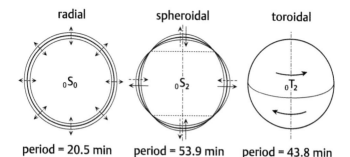

radial	spheroidal	toroidal
$_0S_0$	$_0S_2$	$_0T_2$
period = 20.5 min	period = 53.9 min	period = 43.8 min

6. **The fundamental modes of natural oscillation of the Earth and their observed periods.**

twists in one direction while the lower hemisphere twists in the opposite sense. The lowest frequency of each of these vibrations is called its fundamental mode. The fundamental modes have periods of 20–54 minutes and involve the entire Earth. The sonority of a bell depends on it ringing simultaneously with multiple overtones, and, in a like manner, the free oscillations of the Earth also have higher modes of vibration. Analyses of the free oscillations provide valuable constraints for determining the elastic properties and density distribution in the Earth.

Reflection, refraction, and diffraction of body waves

When a seismic wave initially spreads out from a source in a homogeneous medium, it does so uniformly in all directions and its wavefront is a sphere. A radius of the sphere perpendicular to the wavefront is called a seismic ray; it describes the direction of travel of the wavefront. When seismic energy is incident on the interface separating two materials, it sets up vibrations (i.e. new seismic waves) in the material on both sides of the interface. Laws similar to those of optics govern the way rays from P- and S-waves interact with an interface at which the elastic properties and density—and therefore the seismic velocities—change.

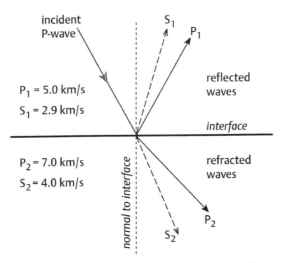

Geophysics

7. **Reflection and refraction of seismic rays set up by an incident P-wave at the interface between media with different seismic velocities. P_1 and P_2 are P-wave speeds of the indicated rays above and below the interface, respectively, and S_1 and S_2 are corresponding S-wave speeds.**

Consider a P-ray incident on an interface, as in Figure 7 (similar considerations apply to an incident S-ray). Above the interface the angles formed between the incident and reflected P-rays and the normal to the interface are equal: this is the law of seismic *reflection*. The ray penetrating the second medium experiences a change of direction that depends on the seismic velocities of the two media. It is called a refracted ray and its direction is determined by a mathematical relationship called the law of seismic *refraction*. For example, if the second medium has a faster speed than the first medium, the refracted ray is bent away from the normal to the interface, so that its direction is shallower; if the second medium is slower than the first, the ray bends towards the normal and becomes steeper. The phenomenon of refraction is familiar from optics; it accounts for the way a straight walking stick inserted in a pool of water appears broken, changing direction at the surface, and explains why the pool appears shallower than it is.

Reflection seismology is an important technique in the detection and description of underground structures on a regional scale, such as the structure of the crust under a mountain range, or on a smaller scale related to local subsurface features that could accommodate mineral or petroleum deposits. The method is based upon the 'echo principle'; it is used to determine the depth of a reflector below a source. When reflection seismology is used to investigate an underground structure, P-waves are created by a sudden or continuous energy source at different locations along a profile across the structure. The waves are reflected at subsurface interfaces where there is a change in the seismic impedance; this is defined as the product obtained by multiplying together the density and the seismic velocity. On returning to the surface the reflections are recorded by arrays of compact portable seismometers called geophones. The prime information gained from a reflection seismic survey consists of the depths to subsurface rock layers as well as the seismic velocities of the layers.

In terrestrial profiling, the source of the P-waves is usually either a controlled explosion or one or more massive vibrating sources. In marine applications, a common source is a specialized air-gun, which discharges a sudden high-pressure air bubble into the water. Several air-guns may be used at a time, linked in an array, and the reflections are detected by arrays of pressure sensors towed at a fixed depth behind an exploration vessel and linked to form streamers that may be several kilometres long.

In addition to the useful record created by the source, a seismogram contains unwanted vibrations that form a background noise, which can spoil the recording just as traffic noise can disturb a conversation. Reflection seismology applies several corrective measures to improve the signal-to-noise ratio on seismograms. Surface wave noise caused by the source ('ground roll') is reduced by combining the geophones in groups. The travel-times of reflections must be corrected for various distortions, resulting from the geometry of the source, geophone, and reflecting surface. A typical reflection survey requires sophisticated data processing and powerful computation, in order to correctly locate and portray the reflecting surfaces. The technique is a way of obtaining a three-dimensional picture of subsurface structures and their seismic velocity profiles. In the petroleum industry, reflection seismology is the most important geophysical technique in exploration. The detailed analysis of some reflection sections can allow changes in seismic impedance to be interpreted in terms of properties such as porosity and permeability, and can indicate the possible presence of gas and liquids.

Refraction seismology is also used to decipher subsurface structure, making use of the way seismic waves are refracted when their velocity changes. Imagine a simple structure of homogeneous layers of rock in which one layer overlies a layer with faster seismic speed. The rays from a surface explosion impact on the lower interface at a variety of angles. Some are reflected back to the surface and some are refracted into the next layer, in which they make a shallower angle with the boundary (Figure 8). One of the incident rays strikes the boundary in such a way that, when it is refracted into the lower faster layer, it travels along or just below the boundary, in the lower layer. This situation is known as critical refraction and it is determined by the velocities of the two layers. The critically refracted ray moves along the interface at the faster speed of the lower layer. In doing so, it disturbs the overlying layer and sets up P-rays that are critically refracted back to the surface, where they are recorded by geophones. Beyond a certain distance from the source (known as the crossover point) the refracted wave arrives before the direct wave in the top layer, because it travels part of the way at the higher speed of the lower layer. (This is analogous to traversing

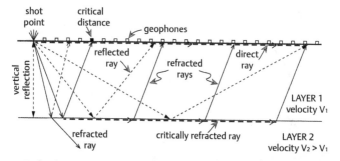

8. Reflection and refraction of seismic ray paths when a shallow layer overlies a layer with higher velocity; velocities are constant in each layer. Geophones on the surface record rays reflected from the interface as well as refracted rays set up by the wave that travels along the interface.

a city by driving a longer distance on a high-speed bypass, which is often faster than taking a shorter route through the city centre.) The travel-times of the first arrivals at geophones in a refraction seismic profile reveal the seismic velocities above and below the refracting interface and its depth.

An additional mechanism that can affect the propagation of a seismic wave is *diffraction*. This occurs when the wave encounters an abrupt or discontinuous surface, such as an intruding body or the edge of a layer. In the same way that ocean waves bend around the end of a solid jetty, the seismic waves are bent around the edge of an obstacle in their path. P- and S-waves diffracted at the core–mantle boundary provide additional data about the Earth's deep structure.

Paths of seismic body waves through the Earth

Earthquake seismologists have deciphered the internal structure of the Earth by studying the reflections and refractions of seismic body waves and the dispersion of surface waves. Imagine a layered Earth in which each layer has a faster seismic velocity than the one above. At each interface, the increase of velocity causes an

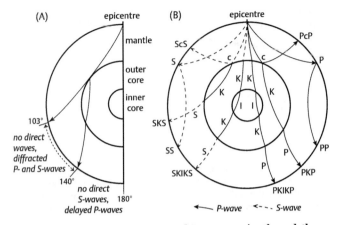

9. (A) Shadow zones for seismic P- and S-waves passing through the Earth. (B) Some examples of refracted and reflected P- and S-waves within the Earth and their relationships to the boundaries of the inner and outer core. The labels PP, PKP, SKS, etc. identify the paths followed by multiply reflected and refracted rays.

Geophysics

incident seismic wave to be refracted away from the normal to the interface to ever-shallower directions (Figure 7). When the angle of incidence on a layer reaches a critical angle, the wave has reached its maximum depth; the wave travels from the interface back towards the surface, refracting towards the normal at each interface it crosses. If each layer is extremely thin, so that the velocity increases smoothly with depth, the path of the wave becomes a curved ray. Within each of the Earth's major internal regions seismic velocities increase progressively with depth. This causes seismic rays to have a curved shape (Figure 9).

The boundaries between the major internal regions are marked by abrupt changes in density and in the velocities of P- and S-waves, resulting in strong refractions and reflections. The seismic arrivals from internal boundaries are identified on seismograms recorded at observatories distributed over the Earth's surface. The distance of a point on the surface from the epicentre of an earthquake—its epicentral distance—can be measured in kilometres or miles along the surface. However, an epicentral distance is more conveniently measured by the angle which the surface segment subtends at the centre of the Earth. Early in the 20th century seismologists observed that no S-waves arrived directly at angular epicentral distances beyond about 103 degrees from an earthquake (Figure 9A). This 'shadow zone' was recognized as evidence that the Earth has a fluid core, because S-waves cannot travel through a fluid. The solid shell surrounding the core was named the mantle. Between 103 and 140 degrees there are also no P-wave arrivals that have directly traversed the core. This is because P-waves that enter the fluid core are refracted at its surface towards the radial (vertical) direction, emerging at epicentral distances beyond 140 degrees. This implies that the P-wave velocity is reduced in the fluid core. This is supported by the observation that P-waves beyond 140 degrees arrive later than expected for an Earth model that assumes a continuous increase of velocity with depth. The velocity of a P-wave depends on both the bulk modulus and shear modulus of a medium in an additive form, and hence the absence of shear slows the P-wave as it traverses the fluid part of the core.

The shadow zones are not devoid of seismic body waves. Reflected and refracted P- and S-waves penetrate the zones, sometimes following complicated paths (Figure 9B). Also, the edges of the shadow zone are not sharp: P- and S-waves that graze the surface of the core are diffracted into the shadow zone as far as 140 degrees. To identify the arrivals of individual rays on a seismogram, seismologists developed a practical labelling scheme. For example, a P-wave arrival that was reflected once at the surface of the Earth is labelled PP, because each segment of the path was a P-wave. Each seismic wave falling upon an interface can generate both refracted and reflected P- and S-waves (Figure 7). An S-wave crossing the core–mantle boundary can be converted partly to a P-wave; this energy can traverse the core, and on re-entering the mantle part of it can convert back to an S-wave; this complex path is labelled SKS, where the letter K represents the P-wave in the core. The letter 'I' is used to identify the segment of a seismic ray that traverses the solid inner core as a P-wave and the letter 'i' identifies a ray that reflects from its surface. For example, PKIKP refers to a P-wave that traversed the entire Earth as a P-wave, whereas SKiKS travelled as an S-wave in the mantle, as a P-wave in the fluid core, and was reflected at the surface of the inner core, repeating the modes of travel on its return path to the surface.

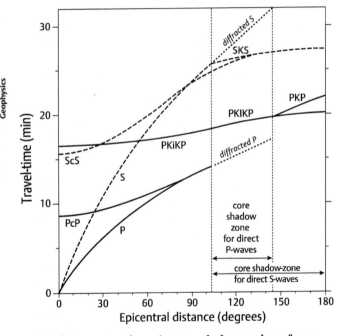

10. **Travel-time curves of some important body wave phases for an earthquake with a focus at the surface of a spherically symmetric Earth. The dotted P-wave arrivals in the shadow zone are diffractions at the surface of the core.**

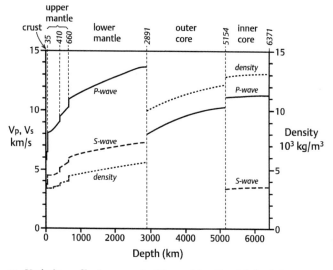

11. **Variations of body wave velocities and density with depth in the Earth's interior.**

Correctly identifying the arrivals on a seismogram and modelling the paths they followed has enabled seismologists to develop a global model of the Earth's internal structure and its composition. Plots of the travel-times of different seismic phases against the epicentral distances from numerous earthquakes yielded globally averaged travel-time curves (Figure 10). Mathematical procedures—called forward modelling and inversion, respectively—are employed to convert these observations at the Earth's surface into the distributions of seismic velocities with depth (Figure 11). The seismic velocities depend on density as well as on the elastic moduli, and so mathematical processing of the velocities yields the variation of density with depth. The analysis of the Earth's natural oscillations, which depend on the radial variations of density and elastic properties, also yields a density model. The successful model is checked by several criteria. Integrated over the Earth's volume the density model must yield the correct mass, and it must also give the correct moment of inertia about the rotation axis. Density changes abruptly in the Earth at the major internal boundaries (Figure 11), indicating that the discontinuities represent changes in composition of the materials on either side of a boundary. However, some internal discontinuities are due to phase changes, in which the composition remains the same but the material structure changes to a form with different density.

Internal structure of the Earth

The observation of seismic waves that passed through the Earth, together with their interpreted reflection and refraction at internal boundaries, revealed the internal structure of core, mantle, and crust. In 1906 Richard Oldham observed that P-waves were delayed as they traversed the Earth and inferred the existence of a central core with reduced velocity. The existence of the core shadow zone for P-waves was established in 1914 by Beno Gutenberg, who calculated a remarkably accurate depth of 2,900 km for the core–mantle boundary, now also known as the Gutenberg discontinuity. Gutenberg also predicted the existence of PcP and ScS waves, where the lower case 'c' indicates that these waves are reflected from the core–mantle boundary (Figure 9B). These reflections were only identified on seismograms much later. In 1926 Harold Jeffreys deduced from the absence of direct S-waves beyond 103 degrees that the core was fluid. In 1936 a Danish seismologist, Inge Lehmann, found evidence for a solid inner core within the fluid core. With a radius of 1,217 km the inner core is smaller than the Moon.

The inner and outer core are extremely hot, with estimated temperatures around 4,000 K (4,273°C) and 5,500 K at the core–mantle and inner–outer core boundaries, respectively. They are believed to have similar compositions, consisting mainly of iron and a small amount of nickel. The enormous pressure keeps the inner core solid, but the outer core is fluid and is consequently in a state of thermal convection. Like a pan of water heated from below the heated fluid becomes less dense and its buoyancy causes it to rise. The viscosity (or 'stickiness') of the fluid in the outer core is low, probably similar to that of water at the Earth's surface. The iron–nickel components of the outer core are constantly solidifying at the inner–outer core boundary, leaving behind lighter elements within the liquid. These rise through the denser outer core, creating a compositional form of convection that supplements the thermal convection. Consequently, the outer core is in a state of turbulent motion. Seismic data from the inner core show that it is not only solid but that it is also anisotropic, that is, the seismic velocity is different in the north–south direction from in the east–west direction. The solid inner core is thought to have a crystalline structure and the observed anisotropy may be due to preferential alignment of the crystals.

In 1909 Andrija Mohorovičič, a Serbian seismologist, observed that the direct P- and S-waves from an earthquake were overtaken by faster waves beyond a particular distance from the epicentre. He interpreted the faster arrivals as waves refracted at the top of a deeper layer with higher seismic speeds (as in Figure 8). This defines the boundary between the crust and mantle, which is now called the Mohorovičič discontinuity, or Moho for short. The Moho-depth is equivalent to the crustal thickness. It averages 22 km for the Earth as a whole but is very variable; oceanic crust is only 5–10 km thick, whereas continental crust is 30–70 km thick and has its maximum thickness under some mountain ranges.

The Earth's main internal divisions form a set of concentric shells (Figure 12). The uppermost mantle has a high rigidity and in combination with the crust constitutes the lithosphere. The outer shell of this composite layer is brittle; it reacts to high stress by fracturing. With increasing depth and temperature the lithosphere becomes ductile; this means it can deform without rupturing, as a metal does when it is drawn out into a wire. At

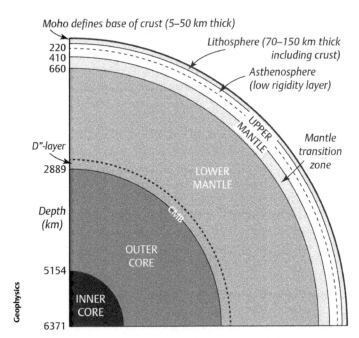

12. Meridian cross-section through the Earth's interior, showing the layered structure and the depths of the principal internal boundaries. CMB is the core–mantle boundary.

oceanic ridges the lithosphere is thin, but the thickness increases progressively away from the ridges to about 80–100 km, while continental lithosphere may be up to 50 km thicker. Geographically, the lithosphere is divided into large plates, whose horizontal extent can be thousands of kilometres. The comparatively thin lithospheric plates move across the Earth's surface, colliding with and separating from their neighbours. These interactions are responsible for tectonic processes on the Earth. The mobility of the plates is facilitated by a layer, called the asthenosphere, which underlies the lithosphere. The top part of the asthenosphere corresponds to a seismic low-velocity zone, on the order of 100 km thick, in which the material is less rigid than in the lithosphere.

The asthenosphere is mechanically weak, deforming in a ductile manner at rates of a few cm/yr. The base of the asthenosphere is not sharply defined but merges into the underlying upper mantle.

The depths between 410 and 660 km constitute the mantle transition zone. In these depths the increasing pressure causes the crystal structure of minerals to collapse to form denser structures. The most common rock in the upper mantle is peridotite, which contains a high proportion of the mineral olivine. A mineral phase change occurs at 410 km depth, where the high pressure rearranges the atoms in the olivine to form a denser structure typical of the mineral spinel. Another phase change occurs at 660 km depth, where the pressure causes the spinel-type structure to convert to an even denser structure typical of the mineral perovskite. This mineral is a magnesium silicate that is thought to typify the composition of the lower mantle. Yet another phase transition has been observed under conditions equivalent to depths of around 2,600 km, involving a phase change from perovskite to a mineral phase called post-perovskite. This is about the depth where seismic evidence indicates an anomalous 150–200 km thick layer of the lower mantle just above the core–mantle boundary. Labelled by historic convention as the D″ layer, this layer is characterized by a variable thickness and variable seismic velocities. Seismic shear wave velocities are anisotropic in this layer, that is, they vary with direction. The velocities of shear waves that vibrate in the vertical plane are a few per cent slower than those that vibrate horizontally, for reasons that are not fully understood.

Seismic tomography

The travel-time curves for P- and S-waves in Figure 10 are global averages. They are based on countless arrival times, recorded by thousands of seismographs at a worldwide distribution of seismic stations, in some cases for decades. In reality, the internal composition and temperature are more complicated than in the idealized, layered Earth. The observed travel-times of individual rays often show small deviations from those in Figure 10. The deviations, called travel-time residuals, may be due to local or regional variations in the velocity structure. The travel-time residuals are employed in seismic tomography to obtain detailed knowledge of the Earth's internal velocity structure. The technique is akin to the medical procedure in which X-rays are employed to image cross-sections of the human body in great detail. Seismic ray paths that pass through a region in many different directions are processed in a tomographic analysis, requiring powerful computers and sophisticated computational techniques.

Seismic tomography can be based on the travel-times of P-waves, S-waves, or surface waves. The travel-time residuals are effectively converted to velocity differences, resulting in three-dimensional images of the velocity distribution. This may then be interpreted in terms of variations in temperature, composition, or rigidity in the region traversed by the seismic rays, assuming that the speed of a ray does not depend on its direction. If a region along a ray path is warmer than its surroundings, seismic velocities may be reduced; if it is cooler, the velocities may be increased. The velocity distribution then reflects the temperature distribution.

Velocity anomalies in the upper mantle have been related by seismic tomography to important aspects of plate tectonics. For example, when the edge of a tectonic plate forms at an oceanic spreading centre, it is hot. By the time, millions of years later, it reaches a boundary where it converges with another plate it has cooled down and its density has increased. The denser oceanic plate is pulled by its own weight—it subducts—beneath the less dense plate, forming a subduction zone, where it sinks into the mantle. The subducting plate is colder and denser than the upper mantle into which it is plunging. The process results in positive velocity anomalies of a few per cent in the upper mantle. The results are usually displayed to advantage in coloured images

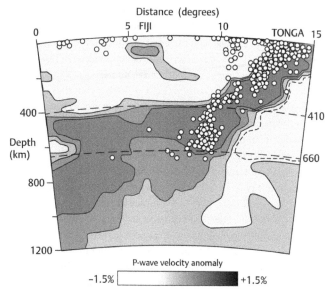

13. Simplified P-wave tomographic section through the Tonga–Fiji subduction zone. Small circles represent earthquake foci; the dark shaded region is the subducting Pacific plate.

of the tomographic sections. A simplified tomographic image of the Tonga–Fiji subduction zone shows the cold Pacific plate bending downwards into the warmer mantle (Figure 13). In this example the subducting plate is deflected when it encounters the 660 km discontinuity, but eventually penetrates into the lower mantle.

Structure of the continental crust

Seismic refraction profiles parallel and perpendicular to the length of a mountain range can be used to investigate crustal structure by determining the depths of refracting rock layers and the seismic velocities in these layers. However, a more detailed picture of the subsurface is obtained from reflection profiles.

Both refraction and reflection investigations have made important contributions to understanding the structure of the continental lithosphere and the depth of the crust–mantle boundary. This kind of large-scale investigation requires cooperation between many partners and input from other disciplines.

In the Canadian Lithoprobe Project, refraction and reflection seismic experiments were coordinated with geological, geochemical, and other geophysical studies to investigate the crustal and lithospheric structure under a 6,000 km long profile across southern Canada at 45–55° N. Extending from the Juan de Fuca active plate in the west to the passive Atlantic margin in the east, the profile documents a complex tectonic history in which orogenic belts have been stacked upon each other. (An orogenic belt forms when a collision between tectonic plates deforms an oceanic or continental plate to form a mountain range.) The Lithoprobe profile revealed a remarkably consistent Moho-depth of 30–40 km, and a variably thick lithosphere ranging from 70 km at the coastal margins to 200 to 250 km in the central continent.

Seismic refraction and reflection formed the basis of the European Geotraverse Project, a similar international cooperation of earth scientists from many disciplines. The purpose was to determine crustal and lithospheric structure along a north–south profile from northern Scandinavia to the southern Mediterranean. This spanned several tectonic provinces from the ancient Precambrian Scandinavian shield to currently active collision areas in the Mediterranean. The project contributed to understanding how the crustal structure across Europe has evolved, and how it continues to do so.

Seismic noise

The high sensitivity of seismometers enables them to detect continuous small ground motions, known as microseisms, that are not caused by earthquakes. After the arrival of P- and S-waves from an active source or earthquake the ground continues to vibrate. This is the result of multiple reflections and diffractions in the inhomogeneous crust, which have the effect of adding noise to the signal. The seismic noise on a seismogram is composed of many frequencies because it has many causes. It has long been known that the interaction of ocean waves on a coastline is one of the causes. Among the other sources are environmental features such as slow-moving landslides or the motion of material in glaciers and rivers, tectonic effects that produce cracking in rocks, and human activity related to urbanization or exploitation of resources.

Seismologists have learned how to use the noise as a passive seismic source that provides a continuous record at a seismometer. If it is recorded for a long enough time, seismic noise can be assumed to reach the seismometer equally from all directions and can be analysed for information about crustal and upper mantle structure. This is achieved by comparing seismic recordings from pairs of seismographs using cross-correlation—a statistical technique in which one record is digitally moved past the other to locate sections of the records that are similar. The combined signals of the two records cancel each other except where they have travelled along a common path, which is the path that joins the recording stations. Each station thus sees the other as an apparent seismic source. The output of the cross-correlation has the characteristics of a Rayleigh surface wave that travelled between the stations, and can be analysed to obtain information about the velocity–depth structure in the crust and upper mantle beneath the stations. With long records from a large number of stations, seismic noise analysis gives enough data to produce a seismic tomographic image of subsurface structure. The continuous nature of seismic noise makes it a viable tool for long-term monitoring of time-dependent environmental changes in a wide variety of situations, such as monitoring slopes endangered by landslides, or recording the motion of glaciers.

The analysis of seismic noise on the Moon is an intriguing example of this passive method. In 1972 the US Apollo 17 mission installed four seismometers on the Moon's surface, forming a triangular array with sides approximately 100 m in length. The network operated until 1977. Cross-correlation of seismic noise recorded continuously at pairs of seismometers for thirty-six weeks in 1976 and 1977 revealed a signal with the characteristics of a Rayleigh surface wave. Analysis of the Rayleigh wave yielded an estimate of the near-surface variation of shear wave velocity with depth and provided new information on the thickness of the lunar regolith (i.e. the layer of loose material covering the surface). Moreover, the signal-to-noise ratio of the derived Rayleigh wave varied with a period of 29.5 days, equivalent to the length of the lunar day. The rhythmic variation was attributed to solar heating, which causes the surface temperature to range from –170°C to +110°C. The high temperature gradient results in cracking and generates the Rayleigh wave.

Chapter 4
Seismicity—the restless Earth

Earthquakes

We were almost asleep when we felt a sudden jolt, closely
followed by a dull rumble lasting a few seconds, as if a heavy
truck had passed by our hotel. We sat up, wide awake, and my
wife exclaimed, 'That was an earthquake!' How could it be?
After all, we were not in California, where seismic events are
common, but in a small village near Birmingham in central
England. Yet it was indeed a small earthquake of magnitude
4.7, and it happened about 30 km away at a depth of 14 km.
It was unusual, but not a rare event.

Hundreds of thousands of earthquakes occur worldwide each
year. Most of them are not noticed by people, because their effects
are so weak that only sensitive instruments record them, or they
occur well away from population centres. However, some of the
tremors are large enough to cause damage, and a few are very
destructive. No place on the Earth can be assumed to be
completely earthquake-free, but most earthquakes happen
in well-defined, relatively narrow seismic zones. More than
90 per cent of earthquakes have a tectonic origin. That is, they
happen on faults, which are planes on which a volume of solid
rock has fractured and blocks of rock on opposite sides of the
fault have moved relative to each other; the fracture surface is

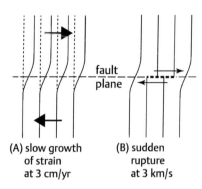

14. Elastic rebound model illustrating the origin of earthquakes on a tectonic fault. Originally straight features (vertical dashed lines) are slowly deformed by movement on the fault until rupture occurs. Long horizontal dashes indicate the line of the fault; short, heavy dashes mark the region of rupture.

called the fault plane. Faulting may occur on more than a single surface, in which case the term fault zone is used.

The elastic rebound model, formulated in 1906 following the San Francisco earthquake, explains how an earthquake occurs (Figure 14). An elastic material is one that reacts to a force by deforming in such a way that it recovers its original shape when the force is removed. However, just as for an over-stretched spring, when the force becomes too large, the elastic limit of the material is reached and the material breaks irreversibly. Although it is based on observations of the San Andreas Fault in California, the elastic rebound model is applicable to earthquakes on other faults. Tectonic forces slowly produce strain in the rocks on opposite sides of the fault at a rate of a few centimetres per year, except on the fault plane itself, where the rocks are locked in contact with each other (Figure 14A). This slow accumulation of strain may continue for many years, even for hundreds or thousands of years, until, at a particular place on the fault, the rocks reach their elastic limit and break, rebounding as a released spring does. The pent-up elastic strain energy is then set free suddenly and violently as an earthquake (Figure 14B). Its energy propagates at speeds of several kilometres per second away from the point of rupture, which is called the focus of the earthquake.

It is important to be able to measure the size of an earthquake. Two properties of an earthquake are in common use for this purpose: its magnitude and its intensity.

Earthquake size: magnitude

The magnitude of an earthquake is a measure of its size based on the energy released by the rupture. Historically, magnitudes have been classified on the Richter scale, which was developed in the 1930s by the seismologist Charles Richter for classifying the sizes of local earthquakes in California. It was later extended to more distant sources. A seismogram contains many frequencies of ground motion, which can be separated by frequency analysis; the lowest frequencies (longest periods) are due to surface waves. The earthquake magnitudes given by the Richter scale were typically computed from surface wave amplitudes at periods of 18–22 seconds. However, very large earthquakes can set up surface waves that carry energy at much longer periods, up to 200 seconds. Consequently, the Richter scale underestimates the size of very large earthquakes with magnitudes greater than about 8.5.

Although treated as a point source, the focus is in reality a ruptured area on a fault plane. Three quantities determine the size of the earthquake: the area of the ruptured section of the fault, the average amount by which it slipped, and the elastic shear modulus of the rocks that ruptured. When these three quantities are multiplied together, the product is called the seismic moment of the earthquake. Fast, modern computers make it possible to analyse the complete waveform content of a seismogram and to calculate the seismic moment from the long period vibrations. The seismic moment is a property of the source, and seismologists consider it to be the best measure of earthquake size. It is related to the energy of the earthquake and is used to define a 'moment magnitude'. This parameter does not saturate for large events and extends the Richter scale to include strong earthquakes. The surface wave magnitude is still used for describing the sizes of small to moderate earthquakes.

At large distances from an earthquake the amplitudes of P-waves with periods around 1 second can be used in a similar way to surface waves to define a body wave magnitude (m_b) for the size of an earthquake. This method has the disadvantage that it saturates for m_b larger than 6, but it is useful for first estimates of the size of distant earthquakes, and also for discriminating between earthquakes and nuclear tests.

Magnitude scales allow estimates of the amount of energy released in an earthquake. For example, a magnitude 6 earthquake releases about as much energy as 15,000 tonnes of TNT, equivalent to the atomic bomb that devastated Hiroshima. In order to handle the wide range of energies released in earthquakes the magnitude scale is logarithmic. As a result, an increase from one size to the next on the magnitude scale involves a tenfold increase in amplitude and a 32-fold increase in the amount of energy released. Consequently, a difference of 2 in magnitude corresponds to a difference in energy of 32^2 or about 1,000 times. For example, the San Francisco earthquake in 1906 had a magnitude of 7.9; by comparison, the largest ever recorded earthquake (in Chile in 1960) had a magnitude of 9.5 and released about 250 times more energy.

The number of earthquakes that occur globally each year falls off with increasing magnitude. Approximately 1.4 million earthquakes annually have magnitude 2 or larger; of these about 1,500 have magnitude 5 or larger. The number of very damaging earthquakes with magnitude above 7 varies from year to year but has averaged about 15–20 annually since 1900. On average, one earthquake per year has magnitude 8 or greater, although such large events occur at irregular intervals. A magnitude 9 earthquake may release more energy than the cumulative energy of all other earthquakes in the same year.

Earthquake size: intensity

The intensity is a qualitative measure of the size of an earthquake based on observed effects, such as the damage it causes. This depends on local conditions such as the ground stability and the robustness of structures. A given earthquake results in different intensities at different places, depending on the distance from the epicentre and the environment of the observer. Volunteer observers report their observations to a coordinating seismic observatory, where analysts collate them to rank the size of the event that was felt at the location of each observer. A numerical value for the intensity is obtained by comparison with standardized observations on a twelve-part intensity scale, which takes into account the effects of an earthquake on people and buildings, and the degree of damage observed. In North America the Modified Mercalli Intensity Scale is used, named after Giuseppe Mercalli, an early pioneer of the method. The European Macroseismic Scale (EMS-98, Table 2) is used in Europe, but the scales are very similar. An Environmental Seismic Intensity Scale (ESI-2007) has been developed to assess intensities of very large earthquakes in uninhabited regions, using environmental effects resulting from the earthquake. Each method enables the estimation of intensity for an earthquake that occurs at the present time and can be applied to past events that are described in old documents and archives, thus providing a means of assessing the historical seismicity of a region. The estimated intensities can be mapped according to their location, and isoseismal maps are created, in which contoured values outline regions of equal intensity around the epicentre. Intensity data are qualitative but they provide a systematic method to describe the overall damage of an earthquake. They represent the only way to assess the past seismicity of a region and are thus valuable for estimating the seismic hazard posed by earthquakes.

Table 2. Abridged version of the European Macroseismic Scale (EMS-98) for earthquake intensity

Intensity	Description of effects
I–IV Light to moderate earthquakes	
I	**Not felt.**
II	**Scarcely felt:** Felt only by a few individual people at rest in houses.
III	**Weak**. Vibration felt indoors by a few people. People at rest feel slight trembling.
IV	**Largely observed:** Vibration felt indoors by many people; outdoors by very few. A few people are awakened. Windows, doors, and dishes rattle.
V–VIII Moderate to severe earthquakes	
V	**Strong:** Felt indoors by most people, outdoors by a few. Many sleeping people awake. A few are frightened. Buildings tremble throughout. Hanging objects swing considerably. Small objects are shifted. Doors and windows swing open or shut.
VI	**Slightly damaging:** Many people are frightened and run outdoors. Some objects fall. Slight non-structural damage to buildings, like hair-line cracks. Small pieces of plaster fall.
VII	**Damaging:** Most people are frightened and run outdoors. Furniture is shifted and objects fall from shelves. Many buildings suffer slight to moderate damage. Cracks in walls, parts of chimneys collapse.
VIII	**Heavily damaging:** Many people find it difficult to stand. Many to most buildings suffer damage, large cracks in walls. Weak structures may collapse.
IX–XII Severe to destructive earthquakes	
IX	**Destructive:** General panic. Many ordinary buildings collapse or show very heavy damage. Monuments topple.
X	**Very destructive:** Many well-built buildings collapse. Landslides occur and cracks appear in the ground.

| XI | **Devastating**: Most buildings collapse, even some with earthquake resistant design. |
| XII | **Completely devastating**: All structures are destroyed. Significant permanent changes to topography. |

Large earthquakes may be preceded by foreshocks, which are lesser events that occur shortly before and in the same region as the main shock. They indicate the build-up of stress that leads to the main rupture. Large earthquakes are also followed by smaller aftershocks on the same fault or near to it; their frequency decreases as time passes, following the main shock. Aftershocks may individually be large enough to have serious consequences for a damaged region, because they can cause already weakened structures to collapse.

Secondary effects of earthquakes

The primary effects of a large earthquake are the result of violent shaking but secondary effects can be equally devastating. The shaking of the San Francisco earthquake in 1906 caused severe damage to structures but it also broke water and gas lines. Consequently, when fire broke out, it could not be brought under control and much of the city burned down. In 1970 a magnitude 7.9 earthquake off the coast of Peru caused extensive direct damage, but it also set off an ice and rock avalanche in the Andes above the town of Yungay. The ice is thought to have melted to form a lubricating cushion under the rock mass, which descended at a speed of around 300 km/hr on the town and buried it, killing 70,000 people.

An oceanic earthquake with magnitude larger than 7.5 may thrust the entire overlying water column upwards (or downwards) and trigger the seismic water wave known as a tsunami. This differs from ordinary waves on the surface of a lake, sea, or ocean, which are set up by wind action and affect only the top few metres of water. By contrast, the entire ocean depth is involved in a tsunami, which travels away from the source at a speed equal to the square root of the product of gravity (9.8 m/s^2) and the water depth. Over the open ocean, which on average is 3,700 metres deep, the tsunami therefore travels at speeds of around 190 m/s, or 700 km/h, the speed of a commercial jet, but the small height of the wave (only a few tens of centimetres) makes it hard to detect. However, when it approaches shallower depths the front of the wave slows down relative to the following water-mass. This causes the tsunami to increase in height as it rushes onshore.

The Sumatra–Andaman earthquake in 2004 had a magnitude of 9.1–9.3 and was thus one of the largest ever recorded. It caused enormous direct devastation and also created the most damaging tsunami in recorded history. About 230,000 people died in this disaster, many of them killed far from the epicentre by the tsunami. The earthquake occurred at a relatively shallow depth of 30 km, on the boundary between the India–Australia plate and the Burma and Sunda portions of the Eurasian plate. A segment of fault measuring around 1,200 km in length slipped by up to 10–15 metres. The initial displacement resulted in a sudden uplift of the ocean bottom by several metres and created a tsunami. It had an amplitude of only 60 cm over the open ocean (observed by satellite) but the wave height reached up to 30 m when it came ashore. Many lives were lost to the Sumatra tsunami in Sri Lanka and India; two people were drowned in Port Elizabeth, South Africa, 8,000 km away from the source. In a similar disaster, a tsunami from the 2011 Tohoku earthquake overwhelmed the coastal protective defences in Japan, destroyed communities, and damaged the Fukushima nuclear plant. About 90 per cent of the world's earthquakes and 75 per cent of its volcanoes occur in the circum-Pacific belt known as the 'Ring of Fire'. Tsunamis occur often enough in this belt for an early warning system to have been set up around the Pacific Ocean, but such a system did not exist for the Indian Ocean at the time of the Sumatra earthquake.

Epicentre location and global seismicity

The further a seismic station is from the epicentre of an earthquake, the longer is the travel-time of a seismic wave. A plot of the travel-time of a P-wave against the distance from the epicentre is a curved line; a similar curved line is found for the slower S-waves. With increasing epicentral distance the separation of the P-wave and S-wave travel-time curves increases (Figure 10). If the difference between the P-wave and S-wave arrival times for a shallow local earthquake is plotted against the time-axis a straight line is obtained. Its intercept with the time-axis gives the time when the earthquake occurred. Seismic velocities vary with depth in the Earth and thus vary along the path travelled by the waves. The velocity–depth profiles are well known, and so the time difference between the P- and S-wave arrivals can be used to compute the distance from the epicentre to the seismic station. The epicentre must lie on a circle centred on the recording station and with radius equal to the epicentral distance. Applying this logic to each of the seismic stations that register the earthquake gives a family of intersecting circles. The epicentre lies at their common intersection point. Often the circles do not intersect exactly at a point but define a small polygon; its geometric centre is taken to be the epicentre, or this can be calculated by more detailed processing. The polygon originates because seismic waves travel to the seismograph from the focus and not from the epicentre.

In practice, the inhomogeneity and ellipsoidal shape of the Earth make the procedure more complicated. After an initial estimate of the earthquake location, the travel-times from it to the seismic stations are computed and compared with the measured times. The earthquake location is then adjusted and the calculations are repeated; the process is repeated until the discrepancies in travel-times are minimized. The method is called forward modelling; it is an important technique for matching a geological model with measured data.

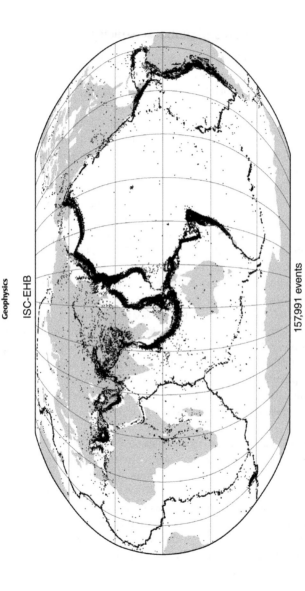

ISC-EHB

157,991 events

15. Distribution of the epicentres of 157,991 earthquakes in the years 1960–2013. The seismicity marks the margins of global tectonic plates. Epicentres located by the International Seismological Centre (ISC) using the EHB algorithm.

The worldwide picture of seismicity is based on the localization of the epicentres and the focal depths of tens of thousands of earthquakes that have been catalogued. The global distribution is not uniform (Figure 15). Most epicentres lie in rather narrow zones, some of which are also characterized by active volcanoes. A zone of shallow earthquakes delineates the oceanic ridges formed by submarine mountain chains. A more diffuse belt extends through the Mediterranean, the Middle East, and Himalaya, and into northern China. The most seismically active parts of the Earth form a prominent belt around the margins of the Pacific Ocean, adjoined by a zone bounding southern Asia. Both shallow and deep earthquakes characterize this seismic belt; in some places the depth distribution of earthquakes extends down into the mantle for up to several hundred kilometres. The seismologists Kiyoo Wadati and Hugo Benioff noted independently that if the earthquakes around the Pacific are plotted on a cross-section normal to the continental margin they define a seismic zone tens of kilometres thick, dipping at an angle of 30–60 degrees and reaching depths of up to 700 km. Initially, the seismic zone was interpreted as evidence for motion on a 'great fault' that dipped below the overriding plate. Subsequently, seismologists found that the observed earthquakes occur as a result of stresses within an inclined slab of lithosphere that subducts—i.e. is pulled down—into the upper mantle in subduction zones.

The worldwide distribution of earthquakes defines the boundaries of the lithospheric plates (Figure 16). In the late 1960s the seismicity pattern delineated the major plates—Pacific, North and South America, Nazca, Eurasia, Africa, Australia–India, and Antarctica—and a few minor plates such as the Caribbean, Cocos, Juan de Fuca, Scotia, and Philippines plates. Subsequently, many smaller plates have been defined, as well as regions of diffuse seismicity that may be incipient plate boundaries. The East African rift zone is a series of rift valleys, thousands of kilometres long, characterized by seismicity and active and dormant volcanoes. Along its length the original African plate is splitting at 6–7 mm/yr

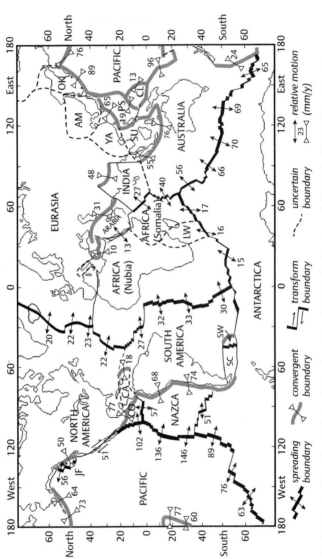

16. **The major lithospheric plates that make up the global tectonic system.** Initials identify the following plates: CA, Caribbean; CO, Cocos; JF, Juan de Fuca; SC, Scotia; SW, Sandwich; LW, Lwandle; OK, Okhotsk; AM, Amur; ~~YA, ... PS, ...; SU, ...; CL, ...; plates with areas less than 10⁶ km² are not shown.~~

into a western Nubia plate and an eastern Somalia plate. Less clearly, analysis of global plate motions suggests that the eastern margin of the former Eurasian plate may consist of four smaller plates, denoted Okhotsk, Amur, Yangtse, and Sundaland. Many lesser plates (not shown in Figure 16) have also been proposed.

The relative motions of the tectonic plates at their margins, together with changes in the state of stress within the plates, are responsible for most of the world's seismicity. Earthquakes occur much more rarely in the geographic interiors of the plates. This is used to justify treating plates as rigid entities in reconstructing their earlier movements. However, large *intraplate* earthquakes do occur, such as the 1811–12 sequence of earthquakes near New Madrid, Missouri, in the middle of the continent. Based upon intensity estimates from historical records the largest shocks in this sequence are thought to have had moment magnitudes in the range 7.5–8. They are the largest known earthquakes to have occurred in the eastern United States. In 2001 an intraplate earthquake with magnitude 7.7 occurred on a previously unknown fault under Gujarat, India, more than 300 km from the nearest plate boundary. It caused extreme damage equivalent to intensity X on the Mercalli scale (Table 2), killing 20,000 people and destroying 400,000 homes.

Fault plane solutions and focal mechanisms

The first arrival of a P-wave on a seismogram may be upward or downward depending on the type of fault and the geographic location of the seismic station relative to the focus. Figure 17 represents a vertical cross-section through the focus of a hypothetical earthquake on an inclined fault plane; its epicentre is at E immediately above the focus. In this case the ground above the plane moves suddenly upward; the same relative motion results when the ground beneath the plane moves downward. The resulting ground motion produces compression in the shaded zones and dilatation (or expansion) in the unshaded zones. As a

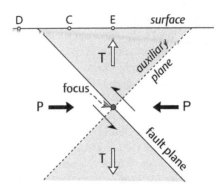

17. Hypothetical vertical cross-section through the fault plane for an earthquake on a reverse fault. Point E on the surface vertically above the focus is the epicentre of the earthquake; the auxiliary plane is perpendicular to the fault plane and intersects it at the focus; P and T are the axes of maximum compressional and extensional strain.

result, the first motion recorded by a seismograph on the surface at C is upwards, while at D the first motion is downwards. The regions of compression (shaded) and dilatation (unshaded) are separated by an auxiliary plane through the focus at right angles to the fault plane. The amplitude of a seismic P-wave within each shaded zone varies with the azimuth of the seismic ray relative to the fault plane. It is zero along the fault plane and auxiliary plane and maximum at 45 degrees on the bisector of the angle between the planes. The direction in which the maximum P-wave first motion is *away from* the source is referred to as the T-axis; it lies in the middle of the (shaded) compressional quadrant. Similarly, the direction in which the maximum first motion is *towards* the source is called the P-axis; it is located in the middle of the (unshaded) dilatational quadrant.

Accordingly, the first motions recorded by seismographs distributed around the focus of an earthquake provide information about the displacement on the fault plane. The known variation of P-wave velocity with depth is used to trace the ray of the first-arrival back to the fault plane, to determine the angle at which it left the focus. This direction is plotted on a stereographic projection, which is a geometric method of converting a spherical distribution to a two-dimensional circular plot. The plot is called a focal mechanism projection, or fault plane solution. The orientations of the fault plane and auxiliary plane plot as arcs, 90 degrees from each other, and the regions of compressional first motion are shaded. The fault plane solution does not in itself identify which is the fault plane and which is the auxiliary plane; this requires separate evidence related to the location of the focus and the surrounding geology.

The type of faulting that occurred can be inferred from the appearance of the focal mechanism diagrams (Figure 18). In a strike-slip fault, the motion is horizontal, so the compressional and extensional quadrants lie in the horizontal plane; the shaded sections divide the circular stereogram symmetrically. In a normal fault, the upper side of the fault slips down relative to the lower side; the shaded regions of compressional first motion are on the rim of the stereogram. In a reverse fault, the upper side moves upwards over the lower side; the region of compression is central and the

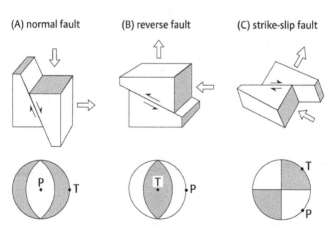

(A) normal fault   (B) reverse fault   (C) strike-slip fault

18. Focal mechanisms and orientations of P- and T-axes for earthquakes on the three main types of tectonic fault.

region of extensional first motion lies on the rim. More complex faulting in which a horizontal fault motion is combined with one of the two vertical fault motions results in a more complicated focal mechanism diagram. Interpretation of the focal mechanisms of earthquakes from the first-motion analysis of seismograms has proved to be an important tool for understanding the type of faulting that occurs during an earthquake and the associated stress regime in the crust and lithosphere. It is the key to understanding the processes that are active at tectonic plate margins.

Earthquakes at plate margins

An example of fault plane solutions on the mid-Atlantic ridge in the Central Atlantic is shown in Figure 19. The first-motion

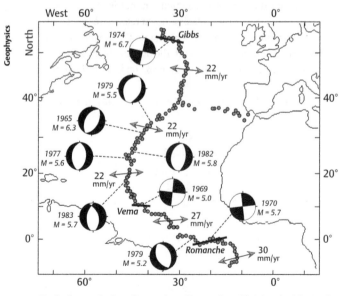

Geophysics

19. **Fault plane solutions for earthquakes on the mid-Atlantic ridge and on the Gibbs, Vema, and Romanche transform faults. Arrows show directions of relative motion with the rates of plate separation in mm/yr.**

patterns show that earthquakes on the ridge axis result from normal faulting, with horizontal T-axes perpendicular to the strike of the ridge. This indicates that tectonic forces are pulling the crust apart at the ridge. Magma from deep in the mantle rises to the surface here, uplifting the oceanic crust to form the ridge and in the process stretching and cracking it. The magma solidifies and builds a continuously renewed edge to the lithospheric plate on either side of the spreading ridge. The ridge process is called sea-floor spreading and the plate margin is called a divergent, or constructive, plate boundary.

Through the Mediterranean and Himalayan regions, and around the Pacific Ocean, the seismicity is related to the presence of mountain belts, deep oceanic trenches, and island arcs. The depth distribution and fault plane solutions of earthquakes indicate that a different tectonic process is active here from at an ocean ridge. A trench marks the place where an oceanic plate is being pulled by its own weight under a less dense adjacent plate. The subduction process forms a convergent, or destructive, plate boundary. This can happen when the adjacent plate consists of continental lithosphere or younger oceanic lithosphere. Earthquake focal mechanisms show the state of stress in the subducting plate.

When an oceanic plate subducts, it is forced to bend downward in order to plunge under the less dense adjacent plate, forming a deep-sea trench. These long, narrow features are the deepest regions of the oceans, in places more than 10 km deep. Focal mechanisms in the shallow part of the subducting plate are of the normal fault type (Figure 18A), indicating that the upper surface of the plate is being stretched. The bending causes horizontal compression under this part of the plate, resulting in earthquakes with a reverse fault mechanism. At the interface between the subducting and overriding plates, a large reverse (or thrust) fault can form (Figure 18B); due to its extreme size it is called a megathrust. All historic earthquakes with magnitude greater than 9 have occurred on megathrusts at subduction zones.

The subducting lithosphere is colder and denser than the underlying asthenosphere and sinks into it. At greater depths the mantle is more rigid than the descending slab and resists its motion, causing a state of compression within the slab, which is evident in the focal mechanisms of intermediate and deep earthquakes.

In addition to divergent and convergent plate boundaries, there is a third boundary type at which the tectonic plates are neither created nor destroyed, but move past each other. This conservative type of boundary joins two segments of a converging or diverging boundary. In 1965 the Canadian geologist John Tuzo Wilson recognized this boundary as a new class of faults related to plate tectonics, and called it a transform fault. Transform faults are associated with long fracture zones in the oceanic crust, which bear witness to past spreading history. A transform fault is the currently active segment of a fracture zone. At first sight it appears to be the same as a transcurrent fault (Figure 20A). However, earthquakes occur along the entire length of a

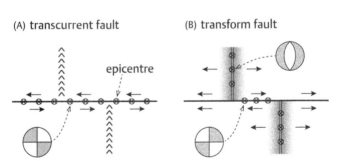

20. Seismicity, fault plane solutions, and relative horizontal motion on strike-slip faults. (A) A transcurrent fault that offsets segments of a linear feature. (B) A transform fault and the adjacent segments of a divergent plate boundary.

transcurrent fault and the offset of two segments of a linear feature increases with time. In contrast, the seismicity on a transform fault—for example at a spreading ridge—is found only on the active zone between the ridge segments (Figure 20B). The distance between the ridge segments does not change and there are no earthquakes on the fracture zone outside the active transform fault. The relative motion on the transform fault is opposite to that on the transcurrent fault, because it is controlled by the directions of spreading on the adjoining ridge segments, as confirmed by the fault plane solutions for both types of fault. Examples of focal mechanisms for transform faults at some fracture zones on the mid-Atlantic ridge are shown in Figure 19. A similar situation can occur on the plate boundary joining subduction zones. For example, the Alpine Fault on the South Island of New Zealand is a transform fault that offsets subduction zones of the Pacific and Indo-Australian plates.

The displacement of a tectonic plate on the Earth's spherical surface is equivalent geometrically to rotating the plate about an axis that passes through the centre of the sphere. Transform faults and their related fracture zones play an important role in plate tectonics, because they can be used to locate the pole of relative rotation of two adjoining plates, known as their Euler pole. Geometrically, the radius of a circle is perpendicular to its circumference and all the radii of a given circle intersect at its centre. Similarly, great circles (meridians) drawn perpendicular to the transform faults on a spreading ridge intersect at the Euler pole of relative rotation of the adjoining plates. Thus the motions of lithospheric plates on the surface of the spherical Earth can be modelled by rotations about Euler poles. Important boundary conditions are placed on the rotations by triple junctions, which are places where three plate boundaries meet. The three types of plate boundary—ridge, trench, and transform fault—can combine to give a large number of possible triple junctions, only a few of which maintain their geometry as the adjoining plates move. An example of a stable triple junction is where the Eurasian, African, and North American plates meet at the Azores in the North Atlantic.

Earthquake monitoring and prediction

In 1996 the Comprehensive Test-Ban Treaty was adopted by a majority of members of the United Nations; a minority of members have not signed or ratified it. Earthquake seismologists play an important role in monitoring compliance with the treaty. A global network of seismic stations records every ground-shaking event. The different characteristics of earthquakes and explosions allow discrimination between them. The source depth provides an initial clue. Earthquakes may have focal depths of tens to hundreds of kilometres; nuclear tests, even in the deepest drill-holes, occur near to the Earth's surface. Secondly, the first motions from an earthquake exhibit a 'beach ball' pattern characteristic of the type of faulting (Figure 18), whereas the first motions resulting from an explosion are all upward on seismograms in every direction around the source. Thirdly, when a P-wave from an explosion impacts the free surface or an internal discontinuity it can generate shear waves (Figure 7), which are eventually recorded as S-waves and surface waves. However, the ratio of the energy carried by short-period (1 s) P-waves to that carried by longer period (20 s) surface waves—i.e. the ratio of body wave magnitude m_b to surface wave magnitude M_s—is larger for explosions than for earthquakes.

In recent years seismologists have discovered that motion on tectonic faults does not always result in a destructive earthquake. A new class of 'slow earthquakes' has been discovered. They result from slow slip on a fault, but the accumulated elastic energy is released slowly instead of abruptly as in an ordinary 'fast' earthquake. The causes of 'slow earthquakes' are not yet understood.

Earthquakes are a serious hazard for populations, their property, and the natural environment. Great effort has been invested in the effort to predict their occurrence, but as yet without general success. Qualitative observations of abnormal animal behaviour and unusual natural phenomena have been interpreted as indicators of impending earthquakes, especially in China, where many regions have a high seismic hazard. However, these precursors cannot be rigorously evaluated and are not useful for prediction. A successful prediction requires estimating reliably where and when an earthquake will occur, and what its size is likely to be. The problem is both scientific and political: the responsible authorities must know fairly precisely when to evacuate an endangered community, because of the logistical difficulties and huge expense entailed.

Scientists have made more progress in assessing the possible location of an earthquake than in predicting the time of its occurrence. Although a damaging event can occur wherever local stress in the crust exceeds the breaking point of underlying rocks, the active seismic belts where this is most likely to happen are narrow and well defined (Figure 15). Unfortunately many densely populated regions and great cities are located in some of the seismically most active regions. The seismic belts are not continuously active and contain gaps where earthquakes have not happened for some time. According to seismic gap theory, these quiet segments of a fault are the most likely locations of future earthquakes. However, this view is not universally accepted by seismologists, because statistical evaluations of the occurrences of earthquakes do not support it.

The slow build-up of tectonic stress on a fault cannot be prevented, but it can be observed and monitored. Sensitive laser-ranging techniques can detect the slow horizontal creep of one side of a fault relative to the other. Satellite geodesy is able to observe motions on dangerous active faults from space. Seismic networks can monitor small earthquakes in order to identify foreshocks and spot any increase in their frequency or magnitude. Despite these scientific and technical advances, it is not yet possible to forecast reliably where or when an earthquake will occur, or how large it is likely to be.

Chapter 5
Gravity and the figure of the Earth

Gravitation, the Earth's shape, and gravity

About 2,500 years ago Pythagoras speculated that the Earth was not flat, as previously believed, but was a sphere. In 240 BC the Greek scholar Eratosthenes estimated the circumference of the sphere by measuring the distance between two places in Egypt where he knew the angle of elevation of the Sun at summer solstice. In 1671 a French astronomer measured accurately the length of a degree of latitude and estimated the Earth's radius to be 6,372 km; a modern value is 6,371 km! Isaac Newton theorized correctly that the Earth's rotation would cause the sphere to be flattened at the rotation poles, and predicted that the flattening would amount to about 0.4 per cent. Eighteenth-century geodetic measurements by French scientists showed that the length of a degree of latitude was shorter near the equator than near the poles, which is what one would observe on a flattened sphere; the estimated flattening was about 0.3 per cent, somewhat less than Newton's prediction. The slight deviation of the Earth's shape—or figure—from a sphere is indeed caused by the planet's rotation and is a very important property. It influences not only the value of gravity anywhere on the Earth, but also how the planet's rotation reacts to the gravitational attraction of extraterrestrial bodies.

When Newton's law of universal gravitation is written as an equation, it defines the gravitational constant G. This important physical constant is extremely difficult to measure, because gravitation is weaker than the other fundamental physical forces. A modern value for G is 6.67408×10^{-11} m^3/kg/s^2. The relative uncertainty in this measurement is 47 parts per million. This is much poorer than most physical constants, which are known at least several thousand times more exactly. However, the product of G and the Earth's mass is known more precisely, to several parts per billion. Likewise, the product of G and the mass of each of the other planets is known precisely. Newtonian gravitation is suitable for resolving situations in most geophysical investigations. However, where the attracting objects are very massive or close to each other, Einstein's theory of general relativity must be invoked; for example, it is needed to explain the orbit of Mercury around the Sun.

The Earth's shape deforms because of the centrifugal force arising from the axial rotation. This force, which is very familiar from our daily experience, is known as an inertial force. Inertia is the resistance of an object to a change in its state of motion. When a car drives round a corner, the inertia of a passenger tries to keep the person moving in a straight line. Contact with the car's interior presses the person towards the centre of curvature of the corner; this force, which keeps the person in the car, is the centripetal force. The equal and opposite force exerted by the person on the car is the centrifugal force. If the car door opens, removing the inwards force, the centrifugal force would eject the person from the car. In the same way, the rotating Earth is subject to a centrifugal force that acts perpendicular to, and away from, the axis of rotation. This force is a tiny fraction (only about 0.3 per cent) of the gravitational attraction, but it has large consequences for the Earth's shape and rotation.

The polar flattening is an elastic deformation resulting from the centrifugal force (Figure 21). It gives the Earth a shape that is

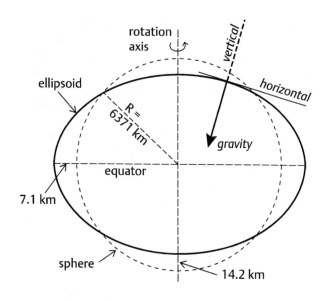

21. Comparison of an ellipsoid of revolution with a sphere of the same volume. The radius, polar, and equatorial differences are shown for the Earth (not to scale). Note that gravity acts in the vertical direction, which is not radial.

symmetrical about the rotation axis; it is an oblate ellipsoid of revolution, or spheroid. This is used by geophysicists as an ideal, mathematically defined shape for the Earth (e.g. for calculating a theoretical value of gravity), and is known as the reference ellipsoid. Its equatorial radius (6,378.14 km) is 21.39 km larger than the polar radius (6,356.75 km); the difference, expressed as a fraction of the equatorial radius, is called the flattening and is equal to 1/298.257 (i.e. about 0.3 per cent). The radius of a sphere with the same volume as the Earth is 6,371.0 km; this is the value used in models of a spherical Earth.

It is often more useful to describe a force in terms of the acceleration it produces. Acceleration is the rate of change of velocity; however, when a force acts on a body with a given mass, acceleration is also the force experienced by each unit of mass. For example, a 100 kg man weighs ten times more than a 10 kg child but each experiences the same gravitational acceleration, which is a property of the Earth. The gravitational and centrifugal accelerations have different directions: gravitational acceleration acts inwards towards the Earth's centre, whereas centrifugal acceleration acts outwards away from the rotation axis. Gravity is the acceleration that results from combining these two accelerations. The direction of gravity defines the local vertical direction (the plumb line) and thereby the horizontal plane. Due to the different directions of its component accelerations, gravity rarely acts radially towards the centre of the Earth; it only does so at the poles and at the equator.

For similar reasons the value of gravity varies with latitude. There are two reasons for this variation. First, at the equator the centrifugal acceleration is maximum and acts outwards, directly opposite to the direction of gravitation, thus reducing the equatorial gravity. Meanwhile, an observer at the poles is located directly on the rotation axis where the centrifugal acceleration is zero, so gravity at the poles is fully equal to the gravitational acceleration. Secondly, the Earth's slightly flattened shape brings the polar surface closer to its centre of mass, strengthening the gravitational attraction. This is partly counteracted by the extra mass beneath the equatorial bulge, which increases slightly the gravitational attraction at the equator. The end result is that gravity is about 0.5 per cent stronger at the poles than at the equator.

Gravity on the Earth's surface and its variation with depth

The instrument used to measure gravity is a gravimeter. The two types in common use are designed to measure either the absolute value of gravity at a point or the change in gravity relative to a reference position, or base station. Absolute gravimeters measure directly the acceleration of a mass while it is falling in an evacuated chamber. A variant of the method consists of a two-way, rise-and-fall measurement, made on a mass that is initially projected upwards. The position of the falling mass is detected with a laser interferometer and the time of fall is measured with an atomic clock. The instrument most commonly used for gravity surveying is a relative gravimeter. In principle this device measures the extension of an elastic spring caused by changes in local gravity relative to a base station, where it is calibrated and an absolute measurement is known. The sensitivity of a relative gravimeter is due to the special design of the spring, which is made of quartz or metal and housed in a temperature-controlled evacuated chamber. It is constructed so as to augment the extension caused by a small change in gravity.

The absolute value of gravity varies with both latitude and altitude. The theoretical value on the reference ellipsoid is known very exactly. It is called normal gravity, as its direction is normal (perpendicular) to the surface of the ellipsoid. A reference formula for normal gravity at any latitude has been derived, based on a precise reference value for gravity at the equator. The geophysical unit for measuring gravity is based on the Gal (named after Galileo), which is defined as 1 cm/s^2. However, this unit is too large to be practical in a geological context. Gravity maps are commonly prepared in units of one thousandth of a Gal, called a milligal (mGal); this is equal to 10^{-3} cm/s^2, roughly one millionth of mean gravity. Gravimeters are extremely sensitive devices and the portable relative gravimeters used in modern geophysical surveying can measure gravity differences a thousand times smaller than a milligal, in the range of a microgal (10^{-6} cm/s^2). A new generation of gravimeter uses superconducting technology to achieve a measurement sensitivity in the range of a nanogal (10^{-9} cm/s^2), equivalent to a trillionth (10^{-12}) of gravity at the Earth's surface.

Using the measured values of gravity and the Earth's radius, in conjunction with the gravitational constant, the mass and volume of the Earth can be obtained. Combining these gives a mean density for the Earth of 5,515 kg/m^3. The average density of surface rocks is only half of this value, which implies that density must increase with depth in the Earth. This was an important discovery for scientists concerned with the size and shape of the Earth in the 18th and early 19th centuries. The variation of density with depth in the layered Earth (Figure 11) was later established from the interpretation of P- and S-wave seismic velocities and the analysis of free oscillations.

A simple calculation for any depth in a spherical Earth shows that the net gravitational attraction of the layers of material above that depth is zero. Gravitational acceleration at any depth in the Earth is therefore due only to the mass of material between that depth and the Earth's centre. This makes it possible to use the density–depth profile to calculate the variation of gravity with depth in the Earth. The calculation shows that gravity at first increases with increasing depth from around 9.8 m/s^2 at the Earth's surface to around 10.7 m/s^2 at the core–mantle boundary, and from there it sinks almost linearly to zero at the Earth's centre.

Pressure increases with depth in the Earth due to the weight of the overlying rock layers and can be computed for a given depth by combining the depth-profiles of density and gravity. In this way the pressure at the Earth's centre has been estimated to be around 360 gigapascals (GPa). A pascal is the physical unit of pressure. Atmospheric pressure at sea level is close to 100,000 Pa, so the pressure at the Earth's centre is about 3.6 million times larger.

The reference ellipsoid and geoid

The energy an object possesses because of its position is its potential energy. The position of a man on a diving board 10 m above a pool gives him a greater potential energy than he would have 5 m above the surface. When the energy is calculated per unit mass, it is called the potential. The mathematical shape of the Earth—the reference ellipsoid—corresponds to a surface on which the potential of gravity is constant; it is called an equipotential surface. The value of the potential on the reference ellipsoid is defined to be the same as that of mean sea level. Importantly, the value of gravity itself is not constant on the equipotential—it is the *potential* of gravity that is constant. The availability of a reference figure makes it possible for geophysicists to calculate the theoretical value of gravity at sea level anywhere on the Earth. The properties of the reference ellipsoid are defined by an international association of geodesists and it is known as the International Reference Ellipsoid.

The reference ellipsoid—bulging over the equivalent sphere by 7 km at the equator and flattened beneath it by 14 km at each pole (Figure 21)—is an idealized representation of the true equipotential surface of gravity, which is called the geoid. The ellipsoid is a very good fit to the geoid but it departs locally from it. The deviations are known as geoid undulations and are commonly of the order of a few metres but can exceed 100 m. They arise for two reasons. First, the ellipsoid assumes that the interior of the Earth is homogeneous, which is not the case. For example, large-scale density differences exist in the mantle due to geodynamic processes. These are the main cause of large, broad geoid undulations. Secondly, the Earth's surface is rough, broken by faults, uplifted in mountains, and mostly covered by oceans that contain deep basins and troughs as well as submarine mountains. Topographic irregularities and density variations cause gravity to differ from the theoretical value and the real equipotential surface to deviate from the ellipsoid. Thus, where gravity is stronger or weaker than expected, the geoid develops bumps or dimples, respectively, relative to the reference ellipsoid.

Prior to the development of satellite geodesy the elevation of the geoid relative to the reference ellipsoid was determined from gravimetric measurements, which required precise levelling. A mathematical procedure devised in 1849 by George Stokes was then used to calculate the geoid height. Beginning in the 1960s, successive satellite missions have provided progressively more accurate determinations of the geoid and the Earth's gravity field. Density anomalies affect the orbits of near-Earth satellites, and as a result it is possible to measure the undulations of the geoid accurately from space (Figure 22). The geoid reaches a depth of 106 m below the ellipsoid over southern India, it is 60 m above the ellipsoid over New Guinea and the North Atlantic, and 40 m above the ellipsoid south of Africa. Generally, the broader a geophysical anomaly (of gravity, the geomagnetic field, or the geoid), the deeper is its source. The large-scale geoid anomalies are too broad to arise from mass anomalies in the crust or lithosphere. They are attributed to anomalous mass distribution in the mantle.

Satellite geodesy

The ability to carry out geodetic and gravimetric observations from satellites has revolutionized geodesy and created a very powerful geophysical tool for observing and measuring dynamic processes on the Earth. Broadly speaking, the various measurement techniques that have been employed fall in two categories: precise location of a position on the Earth, and accurate determination of the geoid and gravitational field.

Many satellites direct radar beams at the Earth's surface and record the reflections. The synthetic aperture radar (SAR) technique, which records and analyses reflections of a radar beam from a swathe perpendicular to the track of the satellite, is able to achieve high resolution imagery of features on the surface. A refinement of the method is Interferometric SAR (InSAR) which combines images obtained on repeated passages over the

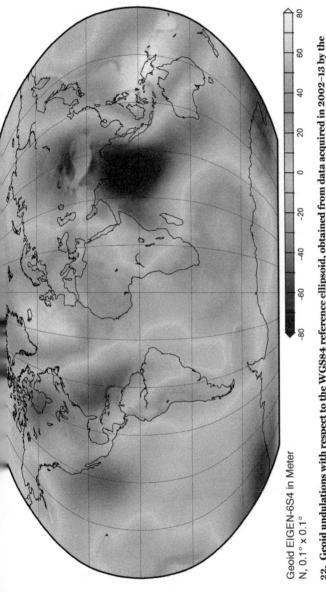

Geoid EIGEN-6S4 in Meter
N, 0.1° x 0.1°

22. Geoid undulations with respect to the WGS84 reference ellipsoid, obtained from data acquired in 2002–13 by the GRACE and GOCE gravity satellite missions.

same region. The radar beam is a waveform, that is, it can be visualized as a sequence of crests and troughs. If any displacement of the surface occurs between the recordings, the radar beams will consist of a different number of wavelengths, or partial wavelengths. If these differ by an odd number of half-wavelengths, one beam will have a crest where the other has a trough and the combined signals will be weakened; this is *destructive* interference. By contrast, if the difference is a number of full wavelengths, crests and troughs will be augmented, which is *constructive* interference. The result of combining the images is a set of interference fringes that emphasize, and allow measurement of, the change in land surface between the satellite over-flights. The method can detect vertical displacements with an accuracy of 1–2 cm that are undetectable by eye, and has a horizontal resolution of around 3 m. Amongst many other applications, InSAR studies have observed the changes produced by earthquakes in the strain fields around faults, and the swelling of volcanoes before eruption and subsidence of the topography afterwards.

The Global Positioning System (GPS) is one of the most important techniques in modern geodesy. It has revolutionized geodetic observations on scales from local and regional surveys to the motions of global tectonic plates. Similar, but separate, systems are operated by the USA (GPS), Europe (Galileo), and Russia (GLONASS). In full operation, each system is designed to consist of 24 satellites in orbits at altitudes of about 19,000–23,000 km (so-called Medium Earth Orbit). The GPS system deploys four satellites in each of six orbital planes inclined at about 55 degrees to the equator and intersecting the equator at positions that are 60 degrees apart in longitude. The satellites emit coded signals in the GHz frequency range that are detected by receivers and converted to a location. Signals from a minimum of four GPS satellites allow the accurate location of a receiver anywhere on the Earth, on land, in aircraft, or at sea, as well as for determining the positions of other satellites. The accuracy of location of a receiver is of the order of a few metres. The Galileo system aims to improve this to better than a metre for general usage, and to centimetre scale for scientific purposes.

Geodesists have developed sophisticated techniques for analysing GPS signals to improve the accuracy of location. One such method is differential GPS. This involves a network of GPS receivers operating in a particular region; the position of one receiver is known very accurately and the other receivers in the network determine their locations relative to it. With differential GPS the accuracy of location improves to a few centimetres. Even higher precision is attained with the technique of carrier-phase tracking, which achieves location accuracy on the millimetre scale. Such exceptional accuracy in determining exactly where a receiver is located has led to various new applications for geodetic surveying. A precise GPS survey of a region can now be made in just a few weeks. The targeted region could be, for example, an area in which there is crustal deformation such as faulting. Repeating the survey at a later time reveals changes in the strain pattern. A network of permanent GPS sites on an active fault can observe motions related to earthquakes on the fault, before, during, and after the event. Such data also provide information about relaxation processes in the lithosphere, such as the rebound of topography that was depressed in the last ice age. Networks of GPS stations provide an independent means of observing actively the present motions and interactions of the global tectonic plates that subdivide the Earth's lithosphere.

Accurate determination of the geoid was an important achievement of satellite altimetry. This technique was initiated soon after the start of space exploration with artificial satellites. The exact altitude of a satellite is measured by tracking its position accurately from a global network of laser stations on the Earth. This gives the height of the satellite above the reference ellipsoid, that is, above mean sea level. Meanwhile, the two-way travel-time of a microwave radar pulse emitted by the satellite measures its height above the sea surface. The difference between the measured heights is the height of the ocean surface relative to the ellipsoid, that is, the geoid undulation. Satellite geodesy has developed with increasing precision and sophistication, as illustrated by two recent space missions.

The Gravity Recovery and Climate Experiment (GRACE) was launched in 2002, deploying two identical satellites in the same polar orbit at an altitude of 500 km. Their horizontal separation of 220 km is monitored accurately with a microwave ranging system that can measure separation changes on the order of 10 micrometres. When the leading satellite approaches a region where gravity is stronger than average, the increased attraction causes it to speed up and, when subsequently it has passed the region, it is slowed down. In each case this changes the distance to the following satellite, which is affected similarly as it passes over the region. The tiny changes in the separation of the satellites are converted into gravity anomalies, resulting in a map of the Earth's gravity field. GRACE results are obtained monthly, which allows observation of changes in the gravity field with time. These include, for example, the gravitational changes that accompany the melting of an ice sheet or that result from the adjustment of the Earth after the ice-age load of a glacier has disappeared. The changes of GRACE data with time have been used to compute the viscosity of the Earth's mantle.

The Gravity Field and Steady-State Ocean Circulation Explorer (GOCE) mission (2009–13) deployed a single satellite in a lower polar orbit, at an altitude of only 255 km. Atmospheric drag at this altitude was compensated by an ion propulsion system. The on-board instrument was a gravity gradiometer, which used three pairs of accelerometers to measure the gravity gradient—the rate at which gravity changes with position—in three mutually perpendicular directions. The height of the ocean surface, determined over many years by satellite altimetry, is made up of two parts. A static part consists of undulations of the geoid due to the anomalous masses inside the Earth. On top of this are superposed the effects of ocean currents, which change with time. When the geoid height is subtracted, the large-scale mean dynamic topography of the ocean surface is obtained. Water flows from high points to lower levels, so the elevations and depressions of the dynamic topography have allowed the ocean current systems to be mapped in unprecedented detail.

The results from these satellite missions are usually presented as coloured maps that dramatically display the spectacular resolution. In regions where data from satellites can be combined with terrestrial data (e.g. gravity measurements on the continents or satellite altimetry over the oceans) the spatial resolution is further improved, so that it is possible to resolve a bump or dent in the geoid or gravitational field smaller than 100 km in horizontal extent.

The tides

The Moon's influence on the Earth's rotation is stronger than that of the Sun or the other planets in the solar system. The centre of mass of the Earth–Moon pair, called the barycentre, lies at about 4,600 km from the Earth's centre—well within the Earth's radius of 6,371 km. The Earth and Moon rotate about this point (like two dancers in a waltz). The elliptical orbit of the Earth about the Sun is in reality the track followed by the barycentre. The rotation of the Earth–Moon pair about their barycentre causes a centrifugal acceleration in the Earth that is directed away from the Moon. The lunar gravitational attraction opposes this and the combined effect is to deform the equipotential surface of the tide and draw it out into the shape of a prolate ellipsoid, resembling a rugby ball. Consequently there is a tidal bulge on the far side of the Earth from the Moon, complementary to the tidal bulge on the near side. The bulges are unequal in size. Each day the Earth rotates under both tidal bulges, so that two unequal tides are experienced; they are resolved into a daily (diurnal) tide and a twice-daily (semi-diurnal) tide. Although we think of the tides as a fluctuation of sea level, they also take place in the solid planet, where they are known as bodily earth tides. These are manifest as displacements of the solid surface by up to 38 cm vertically and 5 cm horizontally.

The Sun also contributes to the tides, creating semi-annual and annual components. Although the Sun's mass is vastly greater than that of the Moon, it is much further away from the Earth and consequently the solar tidal acceleration is only 45 per cent that of the Moon. Imagining each tidal displacement to be shaped like a rugby ball, it is evident that the lunar and solar tides reinforce each other when the 'rugby balls' are parallel. This happens when the Sun and Moon are on opposite sides of the Earth (in opposition), as well as when they are on the same side (in conjunction). These reinforcements occur at full moon and new moon, respectively. In each situation extra high tides result, called spring tides. Midway between these two geometries the 'rugby balls' are perpendicular to each other (in quadrature) and as a result they weaken each other, giving unusually low tides that are called neap tides.

The displacements of fluid and solid mass have a braking effect on the Earth's rotation, slowing it down and gradually increasing the length of the day, currently at a rate of 1.8 milliseconds per century. Evidence for this effect has been deduced from Babylonian, Chinese, Islamic, and European astronomic records covering the past 2,700 years. It can now be measured directly by VLBI, GPS, and other satellite techniques. The reciprocal effect of the Earth's gravitation on the Moon has slowed lunar rotation about its own axis to the extent that the Moon's spin now has the same period as its rotation about the Earth. That is why it always presents the same face to us. Conservation of angular momentum results in a transfer of angular momentum from the Earth to the Moon, which is accomplished by an increase in the Earth–Moon distance of about 3.7 cm/yr (roughly the rate at which fingernails grow), and by a slowing of the Moon's rotation rates about its own axis and about the Earth. In time, all three rotations will be synchronous, with a period of 48 present Earth-days. The Moon will then be stationary over the Earth and both bodies will present the same face to each other. This state already exists between the dwarf planet Pluto and its moon Charon.

Correction and reduction of gravity measurements

Gravity surveying is a valuable technique employed in various types of geophysical investigation, whether on a large scale to analyse global geodynamics, or on a smaller scale in geophysical prospecting for petroleum and minerals. The ability of gravimeters to measure gravity anomalies in the microgal range makes it a useful tool in environmental applications of geophysics, for example in assessing an archaeological site, where buried walls or cavities may cause tiny local changes in gravity.

A gravity anomaly is the deviation of an individual measurement of gravity from the normal gravity at the same location. It is caused by a structure that lies beneath the reference ellipsoid. Gravity is increased over a high-density structure; over a low-density one it is decreased. However, the place where gravity is measured rarely lies directly on the ellipsoid, so various adjustments must be carried out to make the comparison possible. These include local effects particular to the measurement site. First, any topography higher than the gravimeter has a component of attraction that pulls upward against gravity, and weakens a measurement. A nearby valley represents missing material, which also weakens the local gravity. To compensate for these local effects, a terrain correction must be added for hills and valleys surrounding the measurement station. After the effect of topography has been removed, the rough terrain has effectively been replaced by a uniform slab between the gravimeter and the ellipsoid. Next, a correction must be made for this slab. Because it lies outside the reference ellipsoid, it exerts an attraction on the gravimeter that must be calculated and subtracted from the measurement. This is known as a Bouguer correction, and—as with the terrain correction—it requires knowledge of the local rock density. For typical crustal rocks it amounts to around 0.1 mGal per metre of altitude.

After compensating for the attractions of the geological material around and beneath the site, the measurement must be further adjusted for the height of the gravity station above the ellipsoid, because gravity decreases with the inverse square of distance from the centre of the Earth. This 'free-air' effect is the largest correction that must be made for the elevation of the site, amounting to about 0.3 mGal/m. It must be added to a measurement made above sea level and subtracted if made below it. After this sequence of steps, the corrected value of the gravity measurement is as if it had been made on the surface of the ellipsoid. The adjusted measurement can then be compared with the theoretical gravity on the ellipsoid at the latitude of the measurement site.

Note that the Bouguer and free-air corrections are of opposite sign: where the free-air correction is positive, the Bouguer correction is negative, and vice versa. Together, they show a decrease of gravity by 1 mGal for each 5 metres of height above sea level; this emphasizes the need for precise geodetic control of the measurement position. In very detailed gravity surveys a further correction must be made for the effects of the marine tide and bodily earth tide. Their combined effect is about 0.03 mGal, which is much larger than the typical 0.01 mGal sensitivity of a gravimeter.

Further corrections must be made when gravity is measured from a moving platform, such as a ship or aircraft. This is because the eastward component of the vehicle's velocity adds to the speed of the Earth's rotation, increasing the centrifugal acceleration; a westward velocity has the opposite effect. The change can be resolved into a vertical component, called the Eötvös acceleration, and a horizontal component, called the Coriolis acceleration. The accelerations are small but significant and vary with latitude. The Eötvös acceleration acts vertically, directly affecting a measurement of gravity and requiring a compensating correction. The effect is small and proportional to the vehicle's speed. For example, on a ship sailing at 5 knots (~9 km/hr) eastward at 45° N it amounts to around 26 mGal; for an aircraft flying at 200 km/hr the correction is almost 600 mGal. These effects are vastly greater than the gravimeter sensitivity or the size of a possible gravity anomaly, so correcting for them is important.

The Coriolis acceleration acts in the horizontal plane and thus does not affect a gravity measurement. It acts to the right of the direction of motion in the northern hemisphere and to the left in the southern hemisphere. This modifies the horizontal path of any object moving on the surface of the earth. For example, atmospheric pressure causes an air mass to move from a region of high pressure towards one of low pressure. The Coriolis force progressively deflects the direction of the moving air mass until it is geostrophic, that is, moving parallel to the isobars of equal pressure. In the northern hemisphere this causes an anticlockwise wind pattern—a cyclone—around a centre of low pressure; conversely, the air flow away from a centre of high pressure causes a wind pattern in a clockwise sense—an anticyclone. The senses of rotation in a cyclone and anticyclone are inverted in the southern hemisphere.

Bouguer and free-air gravity anomalies

After all corrections have been made, the adjusted gravity measurement is compared to the theoretical value expected at the latitude of the measurement station. The difference is called a *Bouguer* gravity anomaly. If the Bouguer and terrain corrections are omitted, the anomaly expresses only the correction for altitude of the measurement station above the ellipsoid and is called a *free-air* gravity anomaly. Bouguer and free-air anomalies provide important evidence of subsurface density contrasts. Over a structure with higher-than-average density the gravitational attraction is increased; over a low-density structure it is decreased. Rocks have variable densities, even within the same rock type, because of the variability of mineral concentrations. However, typical mean values for the continental crust are 2,300 to 2,500 kg/m^3 for sedimentary rocks, 2,600 to 2,700 kg/m^3 for granite, and 2,400 to 3,000 kg/m^3 for metamorphic rocks. Basaltic and gabbroic rocks in the oceanic crust have higher densities, around 2,800–3,200 kg/m^3. The density contrast (difference) between one rock type and another is small. It may be caused by a structure such as the intrusion of one rock type by another, or it could result from the displacement of rocks by faulting.

The effect of density contrasts is illustrated by the differences between Bouguer anomalies over the oceans and continents (Figure 23). The oceanic crust is comparatively thin, consisting of a few kilometres of water above 5–10 km of basaltic rock. A marine Bouguer anomaly that can be compared with one on the continent is computed by figuratively filling the ocean basin with

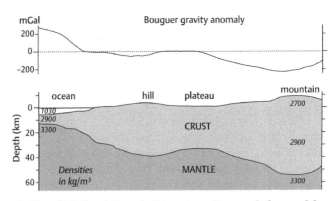

23. Hypothetical variation of a Bouguer gravity anomaly for a model of oceanic and continental crust.

continental rock of typical density 2,670 kg/m^3. The continental crust is about 30 km thick at sea level, therefore upper mantle rocks with a higher density (3,300 kg/m^3) lie closer to the Earth's surface under the oceans than under the continent. The extra mass at shallow depth results in a positive Bouguer anomaly over oceanic crust. In contrast, the high topography of a mountain range is underlain by a thicker crust with the Moho at 60–70 km depth, so crustal rocks with lower density (2,700 kg/m^3) are found in the depths where denser mantle rocks underlie the normal continent. This is referred to as a low-density root of the mountain. It results in a negative Bouguer anomaly over the mountain range.

A free-air gravity anomaly is not corrected for the topography of a continent, nor is it corrected for the variable water depth of an ocean. The free-air anomaly thus responds to both the topography and subsurface density contrasts. For example, over a mountain range the attraction of the topographic mass above sea level is often compensated at least partly by a reduced attraction from a low-density root zone beneath the mountain. Free-air anomalies over mountains that are compensated in this way are quite small. If the mountain has no root zone (or only a small one), the free-air gravity anomaly mirrors the topography. Over the oceans the free-air gravity anomaly maps the bathymetry well, correlating with submarine mountains, seamounts, and trenches. A striking feature is the association of elongate negative free-air anomalies with the deep-sea trenches around the Pacific Ocean. The trenches contain only water or low-density sediment, so there is a strong density contrast with the oceanic crust. Seaward of a trench, the presence of a small positive gravity anomaly attests to the slight upwards bending of the oceanic plate before it subducts beneath the adjacent plate. The flexure is due to the stiffness of the lithospheric plate, which causes it to resist being bent; a sheet of paper, pushed from one end across the edge of a flat desk, bends in a similar way.

Isostasy

Lord Cavendish succeeded in measuring the gravitational constant in 1798. Before the value of this constant was known, it was not possible to calculate the mass or density of the Earth directly from a gravity measurement. In the mid-18th century, scientists attempted to calculate the mean density of the Earth by comparing the acceleration of gravity with the horizontal attraction exerted on a suspended mass by a nearby mountain. This was measured by noting the local deflection of a plumb-line, which defines the vertical direction. To the surprise of the investigators, the deflections for mountains in Peru, Ecuador, and Scotland gave inconsistent and unrealistic values for the Earth's mean density. In the first half of the 19th century, geodetic surveying in the Himalaya also showed anomalous deflections of the vertical near mountains. The unusual results were explained by assigning root-zones to the mountains, in which the rocks have lower density than expected. The attraction due to the visible mountain was thus partly offset by a reduced attraction due to the 'missing' amount of mass in the root. The compensation of topography by a low-density root, such that the mountain effectively floats on an underlying layer, is known as isostasy. Three theoretical models are commonly used to explain the observations.

The first two types, proposed in 1855, are buoyancy models based on the Archimedes principle, which explains how wooden blocks float in water. They envisage a light crust that floats upon a denser substratum. In the Airy–Heiskanen model the crust has a constant density and variable thickness to reflect the topography. It 'floats' in the same way as variously sized blocks of wood in water, which extend to different depths until they have the same pressure at their base. This model has a variable compensation depth, akin to the structure of the crust. In the Pratt–Hayford model the changes in topography are compensated by variable density, with less dense rock under mountains, so that the entire topography has a common compensation depth. This model resembles floating blocks of different woods of different densities, projecting above water to different heights but extending to a common compensation depth. Both isostasy models use a buoyancy mechanism to explain how an extra mass (e.g. a mountain) above sea level is compensated by a lack of mass at depth. The buoyancy mechanism provides local isostatic compensation: the uplift is caused by the local vertical difference in density.

A coin placed on a block of wood floating in water causes the block to sink deeper until a new balance is reached between its weight and the uplift it receives from the water. If the coin is removed, the uplift from the submerged part of the block is too large for equilibrium and the block rises. In the same manner, isostatic compensation causes the crust to move vertically to seek a new hydrostatic equilibrium in response to changes in the load on the crust. Thus, when erosion removes surface material or when an ice-cap melts, the isostatic response is uplift of the mountain. Examples of this uplift are found in northern Canada and Fennoscandia, which were covered by a 1–2 kilometre-thick ice sheet during the last ice age; the surface load depressed the crust in these regions by up to 500 m. The ice age ended about 10,000 years ago, and subsequent postglacial isostatic adjustment has resulted in vertical crustal movements. The land uplift was initially faster than it is today, but it continues at rates of up to 9 mm/yr in Scandinavia and Finland (Figure 24). The phenomenon has been observed for decades by repeated high-precision levelling campaigns. Nordic geodesists now measure the uplift with millimetre-scale accuracy using a network of continuously recording GPS stations.

In a third model of isostasy, formulated in 1931 by Felix Andries Vening Meinesz, the upper layer is an elastic plate that bends downward under a topographic load, but receives support from a 'fluid' substratum. The bending of the plate distributes the load

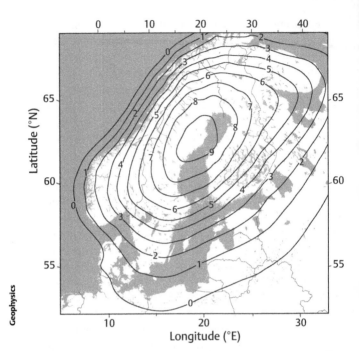

24. Present rates of postglacial land uplift (in mm/yr) in Fennoscandia, derived from repeated precise levelling, tide-gauges, and continuous GPS stations.

over horizontal distances that are wider than the topographic load and results in regional isostatic compensation; this contrasts with the local equilibrium in buoyancy models. The isostatic compensation of oceanic islands and seamounts has been successfully modelled by equating the elastic plate in the Vening Meinesz model with the top part of the lithosphere. However, the thickness of the elastic part of the oceanic lithosphere is noticeably less than its seismic thickness. Evidently the increase of temperature with depth results in anelastic behaviour of the deeper lithosphere. This is the same kind of behaviour that causes attenuation of seismic waves in the Earth (see Chapter 3). It is an irreversible process that depends on the applied stress and also on the rate of change of the stress. This depends on temperature and occurs faster at higher temperatures.

A specific type of anelastic behaviour is viscoelasticity. In this mechanism a material responds to short-duration stresses in the same way that an elastic body does, but over very long time intervals it flows like a sticky viscous fluid. The flow of otherwise solid material in the mantle is understood to be a viscoelastic process. This type of behaviour has been invoked to explain the response of the upper mantle to the loading of northern Canada and Fennoscandia by the ice sheets. In each region the weight of an ice sheet depressed the central area, forcing it down into the mantle. The displaced mantle caused the surrounding land to bulge upward slightly, as a jelly does around a point where it is pressed down. As a result of postglacial relaxation the opposite motion is now happening: the peripheral bulge is sinking while the central region is being uplifted. This behaviour involves both the flexural rigidity of the lithosphere—its resistance to being bent—and the viscosity of the upper mantle. The depth to which the mantle is involved depends on its local structure and on the weight and extent of the surface load.

Chapter 6
The Earth's heat

Heat sources

In popular usage heat and temperature are often taken to be the same, but, although related, they are different concepts which we need to clarify. The molecules of an object are in constant motion and the energy of this motion is called kinetic energy. Temperature is a measure of the average kinetic energy of the molecules in a given volume. It is measured in degrees Celsius (°C) or in kelvins (K), which have the same dimensions but different reference points. Absolute zero on the Kelvin scale, the temperature at which all thermal motion ceases, corresponds to −273.15°C.

The total energy of motion of all the molecules in a volume is its internal energy. When two objects with different temperatures are in contact, they exchange internal energy until they have the same temperature. The energy transferred is the amount of heat exchanged. Thus, if heat is added to an object, its kinetic energy is increased, the motions of individual atoms and molecules speed up, and its temperature rises. Heat is a form of energy and is therefore measured in the standard energy unit, the joule. The expenditure of one joule per second defines a watt, the unit of power. The power consumption of devices in daily usage is measured in kilowatts (i.e. thousands of watts); for example, a modest automobile may have a 100 kW engine.

The rates of heat transfer in the Earth are much smaller than a watt and are measured in milliwatts (mW, i.e. thousandths of a watt). The amount of geothermal heat flowing per second across a unit of surface area of the Earth is called the geothermal flux, or more simply the heat flow. It is measured in mW/m^2.

The Earth's internal heat is its greatest source of energy. It powers global geological processes such as plate tectonics and the generation of the geomagnetic field. The annual amount of heat flowing out of the Earth is more than 100 times greater than the elastic energy released in earthquakes and ten times greater than the loss of kinetic energy as the planet's rotation slows due to tidal friction. Although the solar radiation that falls on the Earth is a much larger source of energy, it is important mainly for its effect on natural processes at or above the Earth's surface. The atmosphere and clouds reflect or absorb about 45 per cent of solar radiation, and the land and ocean surfaces reflect a further 5 per cent and absorb 50 per cent. Almost all of the energy absorbed at the surface and in the clouds and atmosphere is radiated back into space. The solar energy that reaches the surface penetrates only a short distance into the ground, because water and rocks are poor conductors of heat. A rock or beach exposed to the Sun feels warm to the touch, but this is a superficial effect. The daily temperature fluctuation in rocks and sediments sinks to less than 1 per cent of its surface amplitude in a depth of only 1 metre. The annual seasonal change of temperature penetrates some nineteen times deeper, but its effects are barely felt below 20 m. However, glaciations advanced and retreated with a periodicity of up 100,000 yr during the most recent ice age. These long-term temperature changes are able to penetrate the Earth to depths of several kilometres. This has to be taken into account when the heat flowing out of the Earth is measured in wells and boreholes.

The internal heat arises from two sources. Part is produced at the present time by radioactivity in crustal rocks and in the mantle, and part is primordial. Their relative contributions are not known exactly but are thought to be similar. The radiogenic heat source is an ongoing process of heat production by the decay of radioactive isotopes, principally of uranium 238, uranium 235, thorium 232, and potassium 40. These isotopes occur mainly in the crust but some are in the mantle. In contrast, the primordial heat is a remnant left over from the planet's fiery formation. In its early molten state a dense core of heavy elements was formed, surrounded by a silicate mantle of lighter elements, which later became encased in a thin, cool crust. This process of separation is known as differentiation. The Earth subsequently cooled and continues to do so, slowly losing the heat of formation. The residual amount of the original heat is the primordial source of today's internal heat. The heat flowing from the Earth's core into the mantle is thought to be largely primordial.

The internal heat has to find its way out of the Earth. The three basic forms of heat transfer are radiation, conduction, and convection. Heat is also transferred in compositional and phase transitions. Thermal radiation is emitted when the particle motion in an object is converted to electromagnetic waves in the infrared region of the spectrum. However, temperatures are not high enough for this process to be important in the Earth. Heat is transported throughout the interior by conduction, and convection plays an important role in the mantle and fluid outer core.

Heat flow through the Earth's surface

Heat transport by conduction is most important in solid regions of the Earth. Thermal conduction takes place by transferring energy in the vibrations of atoms, or in collisions between molecules, without bodily displacement of the material. The flow of heat through a material by conduction depends on two quantities: the rate at which temperature increases with depth (the temperature gradient), and the material's ability to conduct heat, a physical property known as its thermal conductivity. The product of the temperature gradient and the thermal conductivity defines the heat flow. On the continents the local temperature gradient can be measured by observing the temperatures at several depths in boreholes and deep wells. The thermal conductivity is measured on appropriate rock samples in the laboratory. In the oceans the temperature gradient is measured by plunging a long pipe-like probe into the sediment-covered sea floor. Thermistors (sensitive thermometers) mounted on the surface of the pipe measure the temperature at different depths in the sediment, from which the temperature gradient is calculated. The thermal conductivity of the sediment can be measured *in situ* in the ocean floor, or later in the laboratory on sediment samples if a core is taken when the pipe is withdrawn.

Heat flow varies greatly over the Earth's surface depending on the local geology and tectonic situation. The estimated average heat flow is 92 mW/m^2. Multiplying this value by the Earth's surface area, which is about 510 million km^2, gives a global heat loss of about 47,000 GW (a GW, or gigawatt, is one billion watts). For comparison, the energy production of a large nuclear power plant is about 1 GW.

About one-third of the Earth's heat loss is through the continents. The average heat flow for continental areas is 71 mW/m^2, but the values vary greatly with the type of rock in a given area and its age. The lowest values of 20–40 mW/m^2 are measured over the Precambrian shield areas. These are the oldest regions of crust on the continents. Except for local areas in which radiogenic isotopes are concentrated, the ancient rocks that form these stable regions (called cratons) have had a very long time to cool down since their formation. By contrast, two-thirds of the Earth's heat loss is through the oceanic lithosphere, where the heat flow averages 105 mW/m^2, much higher than the continental heat flow. The oceanic heat flow is highest on oceanic ridges (Figure 25), where the heat flow may locally exceed 500 mW/m^2.

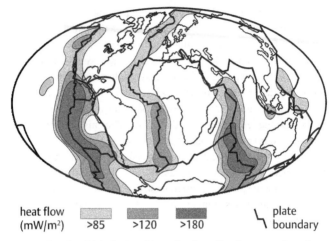

heat flow (mW/m²) >85 >120 >180 — plate boundary

25. Regions in which the Earth's surface heat flow is greater than the global mean value.

Radiogenic minerals are scarce in the oceanic crust, so only a few per cent of the oceanic heat flow has a radiogenic origin.

The high values at oceanic ridges are due to the formation of fresh oceanic lithosphere at active spreading centres. Here, upwelling hot magma brings heat to the surface by the process of advection. Resembling convection, advective heat transport accompanies the movement of a fluid, but this form of heat transfer is not driven by heat directly. For example, a lava flow is caused by gravity, and the expulsion of hot material in a volcanic eruption is due to pressure differences; advective heat is transported by the motion of the lava in each case. By contrast, thermal convection is powered by the buoyancy forces that result from differences in density and temperature.

The oceanic heat flow decreases with distance from a spreading ridge because the lithosphere cools by conduction as it ages and

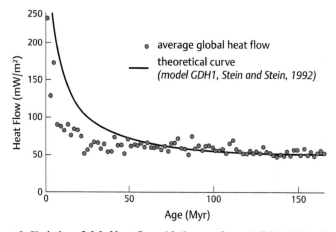

26. Variation of global heat flow with the age of oceanic lithosphere. Points are global averages for a particular lithospheric age. Low values near the ridge axis are attributed to the circulation of hydrothermal fluids that carry away part of the heat.

moves away from the spreading ridge. The cooling is well understood and can be explained theoretically. The oceanic heat flow is found to be inversely proportional to the square root of the age of the lithospheric plate (Figure 26). The age of oceanic crust can be determined from the rate of sea-floor spreading and the distance from a ridge. Near to the ridge, heat is also removed by hydrothermal circulation through cracks and fissures in the young oceanic crust, causing the conductive heat flow to fall below the theoretical curve. As well as cooling with age, the oceanic plate thickens in proportion to the square root of its age. At a ridge axis the young lithosphere is only a few kilometres thick, but the thickness increases as the plate ages, reaching a value of around 100 km where the lithosphere is older than about 60 million years (Myr). Moreover, as the plate ages and cools it becomes denser and subsides. Consequently, the depth of the ocean also increases as the square root of the plate's age. Where an oceanic plate converges with an adjacent plate, a deep trench forms. These are the deepest parts of the world's oceans and are characterized by very low heat flow values.

When a lithospheric plate subducts at a converging plate margin, the leading edge of the cool slab bends downward into the hotter mantle. Thermal conduction is a slow process and for millions of years the interior of the subducting slab remains colder and thus denser than its surroundings. The 'negative buoyancy' causes the slab to sink into the less dense upper mantle. In the process it exerts a pull on the lithospheric plate which is stronger than other forces that act on it. The 'slab pull' is the dominating force that drives plate tectonic motions.

Temperature inside the Earth

Compared to seismic velocities, density, and gravity, the temperature at different depths in the Earth is not well known. Direct measurement is impossible below the upper few kilometres that have been penetrated by boreholes or deep mines. In general, the temperature increases with depth, defining a curve called the geotherm. This is affected by local tectonic conditions as well as by the age of surface rocks. The curve is not known precisely: estimates of the temperature at crucial depths sometimes differ by hundreds of degrees. Near the Earth's surface the temperature gradient averages around 25–30°C per kilometre. A continual increase at this rate would give an unrealistic temperature of about 200,000°C for the centre of the Earth and most of the interior would be molten. Seismic results show that this is not the case, and so the temperature gradient must decrease with depth. It is steepest in the crust, and increases much more slowly in the mantle.

The relationship between the geotherm and the temperature at which melting begins (known as the solidus, or 'softening point') can be inferred from the physical state of the Earth's internal structure deduced from seismic travel-times (Figure 10). The high pressure in the core raises the melting point of iron, which is the core's main constituent, and enables the inner core to be solid despite the high ambient temperature. This implies that the temperature of the inner core must be lower than the melting point of iron under core conditions. On the other hand, the outer core is molten, and so its temperature lies above the melting point of iron. The boundary between the inner and outer cores marks an anchor point for the temperature curve; it is the melting point of iron at the pressure and temperature corresponding to that depth. Outside the core, seismology shows that the mantle and crust are solid for the passage of P- and S-waves, so temperatures in these domains lie below the melting point. A reduction in shear wave velocity in the asthenosphere implies that it has a reduced rigidity and, by inference, that the temperature comes closer to the solidus. Throughout the mantle the difference between the actual temperature and the solidus determines the ability of different parts of the interior to flow over long time intervals. In turn, this influences motions of the lithospheric plates at the Earth's surface and determines the tectonic consequences of their interactions with each other.

The inaccessibility of the Earth's interior means that the temperature–depth profile (Figure 27) must be determined indirectly. This is attained partly from theoretical models based on known or estimated properties of materials composing the mantle and core. An important limitation on the actual temperature–depth curve is the adiabatic temperature profile. An adiabatic thermal process is one in which heat is neither gained nor lost. This can be the case when a process occurs too quickly to allow heat to be exchanged, as in the rapid compressions and expansions during the passage of a seismic wave. The variation of temperature with depth under adiabatic conditions defines the adiabatic temperature gradient. The dependence of the adiabatic gradient on physical properties, such as the coefficient of thermal expansion and the specific heat, is well understood and described by thermodynamic equations. The equations can be applied in

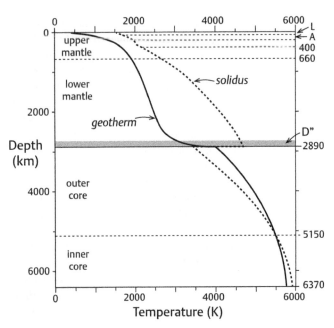

27. Variations of adiabatic and melting-point temperatures with depth in the Earth's interior.

provinces such as the lower mantle or outer core that have consistent physical and thermal properties. The results indicate an adiabatic temperature gradient of around 0.8 K/km in the outer core.

The other limit on the actual temperature profile is the behaviour of the solidus curve with depth. Melting temperatures of appropriate minerals at key depths are obtained from laboratory experiments at high pressures. An important method employs X-ray diffraction to observe the changes in structure and melting point of minerals at extremely high pressures produced by a device known as a diamond anvil cell. The variation of melting point is measured over a range of pressures, and the observations are extrapolated to greater depths outside the experimental range. For example, the melting point of iron at the boundary between the outer and inner core has been estimated to be 6,200 K from experiments carried out at lower pressures and extrapolated to the 330 GPa pressure at the boundary.

Thermal convection and mantle flow

Consider what would happen in a fluid if the temperature increases with depth more rapidly than the adiabatic gradient. If a small parcel of material at a particular depth is moved upward adiabatically to a shallower depth, it experiences a drop in pressure corresponding to the depth difference and a corresponding adiabatic decrease in temperature. However, the decrease is not as large as required by the real temperature gradient, so the adiabatically displaced parcel is now hotter and less dense than its environment. It experiences a buoyant uplift and continues to rise, losing heat and increasing in density until it is in equilibrium with its surroundings. Meanwhile, cooler material adjacent to its original depth fills the vacated place, closing the cycle. This process of heat transport, in which material and heat are transported together, is thermal convection. Eventually the loss of heat by convection brings the real temperature gradient close to the adiabatic gradient. Consequently, a well-mixed, convecting fluid has a temperature profile close to the adiabatic curve. Convection is the principal method of heat transport in the Earth's fluid outer core.

Convection is also an important process of heat transport in the mantle. As already noted, the mantle exhibits viscoelastic behaviour. It reacts like an elastic solid during the rapid passage of seismic waves and responds immediately and reversibly to an applied stress. However, when stressed for long time intervals at high temperature and pressure, the mantle behaves like a highly viscous fluid. In response to an applied stress it exhibits an irreversible increase of strain (i.e. change of shape) with time.

This is due to the migration of crystal defects in the mantle minerals. It is a thermally activated process, in which the strain rate increases rapidly with increasing temperature. Over long time intervals, this enables the mantle to flow like a hot viscous fluid, resulting in a temperature-driven bodily transfer of mass and heat, that is, thermal convection.

It is important to appreciate the timescale on which flow occurs in the mantle. The rate is quite different from the familiar flow of a sticky liquid, such as blood or motor oil, which are three times and 250 times more viscous than water, respectively. The mantle is vastly stiffer. Estimates of viscosity for the lower mantle are around 10^{22} Pa·s (pascal seconds), which is 10^{25} times that of water. This is an enormous factor (similar to the ratio of the mass of the entire Earth to a kilogram mass). The viscosity varies within the mantle, with the upper mantle about 20 times less viscous than the lower mantle. Flow takes place in the mantle by the migration of defects through the otherwise solid material. This is a slow process that produces flow rates on the order of centimetres per year. However, geological processes occur on a very long timescale, spanning tens or hundreds of millions of years. This allows convection to be an important factor in the transport of heat through the mantle.

The flow patterns of mantle convection are influenced by thermal boundary layers, which are layers where the rheology and mechanism of heat transfer change. The cold lithosphere, in which heat transfer takes place by conduction, forms an upper boundary layer. Another is the hot D″-layer above the core–mantle boundary, where the fluid core adjoins the viscoelastic mantle. In this anomalous seismic zone the temperature falls steeply by about 1,400 K between the outer core and the lower mantle. As a result, the layer plays an important role in mantle convection. It influences the heat flow across the core–mantle boundary and thus may affect the processes by which the geomagnetic field is created.

Mantle convection and plumes

Mantle convection plays a crucial role in the cooling history and evolution of the planet. Fresh lithosphere is created at ocean ridges and old lithosphere is consumed in subduction zones; this cycle is maintained by convection throughout the mantle. The deepest earthquakes occur in subduction zones, but few focal depths lie below the 660 km transition zone. Focal mechanisms of deep earthquakes indicate that the sinking plate is being compressed at these depths, implying resistance to subduction beyond the transition zone. However, seismic tomography has shown that cold subducting plates can penetrate the transition zone (see Figure 13), in some cases descending as far as the D″ layer adjacent to the core–mantle boundary. This has led to a model, favoured by most geodynamicists, of whole mantle convection. This process takes place between an upper thermal boundary layer at the lithosphere and a lower thermal boundary at the core–mantle boundary. The entire mantle undergoes convection in the intervening space. The lithosphere participates in the convection, acting as a stiff outer lid. The bending of the lithosphere at subduction zones causes stress and internal deformation, breaking up the lithosphere into individual plates. Computer models of the break-up of the lithosphere into plates have successfully reproduced the size distribution of the present-day plates.

The D″ layer plays an important role in whole mantle convection. It is regarded as the source of mantle plumes. These are visualized as pipe-like conduits that allow relatively fast-flowing, low-viscosity magma to rise rapidly through the more viscous mantle (Figure 28). The narrow plume, perhaps only 100 km in width, is estimated to rise rapidly through the mantle, and in the process it develops a broad head, giving it a cross-section like a mushroom. When it reaches the base of the lithosphere, the plume head flattens out and causes the lithosphere to bulge upward to form an uplift or swell that may be more than 1,000 km wide. Simultaneously it

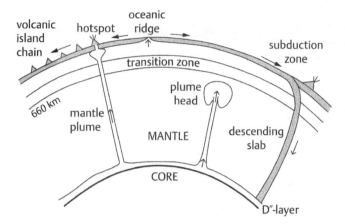

28. Cartoon representation of a cross-section through the mantle, showing some of the features involved in mantle convection.

reduces the depth of the ocean by hundreds of metres and elevates the geoid locally by several metres; both of these features are well documented.

The mantle plumes feed so-called hotspots on the Earth's surface. These are places in the oceans (e.g. Hawaii) and on the continents (e.g. Yellowstone) where there is high heat flow and sustained intraplate volcanism. This is a type of volcanic activity that occurs far from a plate boundary, and which produces basalts with a slightly different chemical composition from those that originate at a spreading ridge. Roughly forty–eighty hotspots have been postulated worldwide, although not all are well documented. A hotspot is anchored to the mantle where its feeder plume is located, creating a volcanic island on an oceanic plate. As the tectonic plate moves over it, the hotspot creates a trace of its motions in the form of a chain of volcanic islands. Many such island chains, such as the Hawaiian–Emperor chain, have been formed in this way. The same behaviour happens on the continents. In North America the Yellowstone hotspot has created a long chain of volcanic features across the state of Idaho and into northern Wyoming.

Mantle plumes are visualized as comparatively thin columns of hot material, on the order of 100–200 km in thickness. Seismology has not been able to establish unequivocally the presence of these narrow features beneath hotspots. In part this is because many hotspots are oceanic and lie far from seismic networks, which are predominantly on land, and this makes it difficult to image a plume. Some seismic evidence that claimed to show their existence in tomographic sections has been questioned on the basis of the imaging method used. On the other hand, seismic tomography has revealed two large, broad regions of low shear wave velocity, several thousand kilometres in extent, located in the lower mantle under Africa and the central Pacific Ocean, respectively. They have been called superplumes; their size and geometry are different from 'mushroom-shaped' plumes. Shear wave velocities are reduced in the superplumes, but their velocity structure is complex and the reduction in shear wave velocities may be due to compositional variation rather than temperature anomalies.

An alternative model of mantle convection is preferred by some scientists, in which a separate circulation takes place above the 660 km transition. This would enable a return flow of material from a subduction zone to a ridge, and requires two convection systems in the mantle. In this model the oceanic island basalts are presumed to result from melting at shallow depths in the upper mantle, with ensuing slight isotopic differences with respect to the oceanic ridge basalts. Stretching and cracking of the lithosphere is then assumed to allow the melt to rise to the surface to produce the hotspot. Although there are fewer adherents to this model, their arguments have fuelled an intense debate.

Chapter 7
The Earth's magnetic field

The geomagnetic dynamo

Our planet is surrounded by a magnetic field, which originates inside the Earth's molten core. The magnetic field is very important for life on the Earth as it acts as a shield against harmful radiation from space, especially from the Sun. For centuries it has helped travellers to navigate safely across uncharted regions. A magnetic compass aligns with the field and points almost—but not exactly—towards true north. Mankind has learned to cope with the small directional deviations and even to interpret them advantageously to locate important metallic resources. Yet to most people magnetism remains a mysterious and little understood phenomenon.

Magnetism has been used for navigation since the 11th century, following the invention of magnetic compasses that used magnetized iron needles to seek the northerly direction. In 1600 William Gilbert showed that the Earth itself resembles a giant magnet, with its centres of magnetic force concentrated near the geographic poles. A freely hanging magnet will align with the Earth's magnetic field, and its opposing ends were recognized as north- or south-seeking poles, eventually shortened to 'magnetic poles'. An understanding of the laws of magnetism emerged, based on the concept of poles. A simple magnetic field with two poles was called a dipole, a field with four poles a quadrupole, eight poles formed an octupole, and so on. A proper physical understanding of electricity and magnetism evolved early in the 19th century, when it became evident that all magnetic fields—on macroscopic, microscopic, and atomic scales—arise from electrical currents. Magnetic poles are a fictitious, but sometimes useful, concept. The close relationship between magnetism and electricity was emphasized by the discovery that a changing magnetic field causes an electrical current to flow in a conductor. In 1872 James Clerk Maxwell quantified the known relationships between electrical and magnetic phenomena in a set of equations, which subsequently led to an understanding of electromagnetic radiation and the interrelationships between electricity, magnetism, and light.

The conditions for generating the Earth's magnetic field are met in the molten core. The core fluid is a good electrical conductor, so the physical flow of the fluid through magnetic fields in the core induces electrical currents, which in turn produce additional magnetic fields in a self-reinforcing process. The fluid flow and electromagnetic interactions between different parts of the core are complex, but massive computing power has made it possible to model what is happening when the magnetic field is generated. The process involves turbulent fluid motions in the outer core. In addition to buoyancy-driven thermal convection, there is a fluid motion called compositional convection, which results from solidification of the inner core. The core fluid consists of iron, nickel, and lower-density elements. When the inner core solidifies, the lower-density elements are left behind in the core fluid. Their lower density makes them buoyant, which causes them to rise and sets up a compositionally driven type of convection.

The motions of the core fluid are further influenced by the Coriolis force that acts on all moving objects on the rotating Earth. This force acts perpendicular to both the rotation axis and the direction of motion. On the Earth's surface, as we have seen, it influences the weather by forcing wind patterns to form cyclones and anticyclones. In the Earth's core its effect is to cause the convecting fluid to form spiralling columns parallel to the rotation axis, thereby determining the geometry of electrical currents in the core. The buoyancy, rotational, and Coriolis forces contribute to a self-sustaining dynamo process that generates the geomagnetic field.

Geomagnetism

The geometry of magnetic field lines is complicated inside the core, but becomes simpler with increasing radial distance from the centre. Mathematical functions, called spherical harmonics, were developed by Carl Friedrich Gauss in the 1830s to provide a detailed description of the geomagnetic field. Using the sparse data available at the time, Gauss showed in 1839 that most of the magnetic field conformed to a dipole and originated inside the Earth. Spherical harmonic analysis is used in modern geophysics to describe the global distributions of magnetic, gravity, thermal, and various other geophysical properties. Applied to geomagnetism it defines the International Geomagnetic Reference Field (IGRF). This is the theoretical magnetic field that best describes the present field at any given latitude and longitude. The IGRF consists of a series of terms that, with increasing number, describe ever smaller, more detailed features of the magnetic field. However, the direction and intensity of the geomagnetic field change slowly with time, a feature known as secular variation (secular in this context refers to the slowness of the changes). Some secular changes occur on a human timescale over months, years, or centuries but others—particularly of the dipole—take place on timescales spanning thousands to millions of years. To cope with the short-term changes the magnetic field is constantly monitored and the coefficients of the IGRF are updated at regular intervals, usually every four years.

The magnetic field at Earth's surface is dominantly that of an inclined dipole (Figure 29). If the best-fitting dipole is subtracted

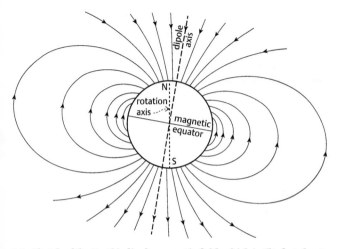

29. Sketch of the Earth's dipole magnetic field, which is tilted at about 10° to the rotation axis. Note that the angle of inclination, at which the field lines cross Earth's surface, gets progressively steeper with increasing latitude.

from the total field, the residual field that is left is referred to as the non-dipole field (NDF). There are two sources for the NDF. The long wavelength components are generated in the molten core by the fluid motions that also generate the dipole. The short wavelength components are due to magnetized rocks in the crust, with sources ranging from several kilometres to hundreds of kilometres in extent.

The magnetic field is measured with an instrument called a magnetometer. Originally designed during the Second World War for anti-submarine warfare, the instruments have subsequently been used widely in geophysical exploration. There are two main types. Scalar magnetometers measure the total intensity of the field; vector magnetometers measure the field component along a known direction. A group of three vector sensors arranged at right angles to each other can measure both the strength and the direction of the field. The unit of measurement of magnetic field is the tesla. This is a very strong field, much larger than the fields encountered in geophysics: a large coil or an electromagnet is needed to produce it. The sensitivity of a magnetometer is better than one nanotesla (nT), which is one billionth of a tesla. This is the practical unit of measurement for the geomagnetic field and its anomalies; advanced magnetometers have a sensitivity of about 0.01 nT. The geomagnetic field has an average intensity of around 50,000 nT.

The magnetometer may be installed at a fixed location, such as a magnetic observatory. A worldwide network of observatories coordinates continuous monitoring of the global magnetic field, recording its fluctuations with time. Alternatively, the magnetometer may be mobile and used for a local, regional, or global investigation. In field surveys at or near the surface a magnetometer may be deployed from a ship or aircraft. Magnetic surveying is relatively inexpensive and fast. Consequently, it is an important technique in applied geophysics, where it is used to explore an unknown area for its potential mineral wealth. It is also used in environmental geophysics and archaeology for local surveys. The goal of any magnetic survey is to identify places where the measured field differs from the theoretical value expected at that place. The deviation, called a magnetic anomaly, is due to inhomogeneity of the subsurface magnetization. Some anomalies may be local, of shallow origin, and may indicate a crustal feature, such as a mineralized dyke or a geological structure of commercial interest. Large regional anomalies have a deep-seated origin.

An important regional magnetic anomaly is found in the South Atlantic. Maps of the geographic distribution of the non-dipole field show a large anomalous region over the Atlantic Ocean to the east of South America, where the NDF has a strongly negative value. The origin of the South Atlantic anomaly is attributed to a region on the core–mantle boundary beneath the anomalous area, where the magnetic field has the opposite direction to the usual magnetic field. This reversed magnetic flux diminishes the entire magnetic field in that region. Magnetic fields affect electrical currents and the Earth's magnetic field lines act as a shield that deflects electrically charged particles from space that continually bombard the planet. The South Atlantic anomaly reduces the shielding effect locally and allows extraterrestrial radiation to come closer to Earth's surface. The elevated radiation dosage can endanger communication satellites and the health of crews and passengers in high-flying aircraft that traverse the region, especially during times of increased solar emissions, when the Sun ejects enormous masses of charged particles.

Since 1980 several space missions (denoted by acronyms like MAGSAT, ØRSTED, CHAMP, SWARM) have been dedicated to measuring the geomagnetic field. These satellite-borne magnetic surveys have greatly improved our knowledge of the field, especially in regions where marine and land surveys previously had large blank areas. An example of this type of satellite-borne magnetic survey is the SWARM mission, launched by the European Space Agency in 2013. It employs a novel constellation of three satellites in near-polar orbits, each carrying both a vector and a scalar magnetometer. Two satellites orbit side by side, separated by about 150 km, at an initial altitude of 460 km; the third satellite orbits above it at about 530 km altitude. Coupling their outputs in pairs enables them to be used as horizontal and vertical magnetic gradiometers, which measure differences in the field from one position to another. This arrangement minimizes daily variations in the geomagnetic field and increases the ability to resolve small-scale magnetic anomalies. In addition to providing better definition of the magnetic fields on the Earth's surface the data from this mission contribute to a more detailed picture of the field on the surface of the core. They also record temporal changes in the field on a short timescale.

The axis of the inclined dipole that best fits the magnetic field is currently tilted at about 10 degrees to the rotation axis. Its poles are located where the dipole axis cuts the Earth's surface. This places the north *geomagnetic pole* at latitude 80.4° N, longitude 76.7° W, in northern Canada, and the south pole at an equivalent position in the southern hemisphere. However, the presence of the NDF means that the total field is not vertical at these poles. The total field is vertical at the *magnetic dip-poles*, which are not symmetrically opposite each other but lie at 86.3° N, 160° W and 64.2° S, 136.5° E, respectively. This means that a compass needle usually does not point to the geographic north pole defined by the rotation axis. As a result, a correction must be made for local deviations when a magnetic compass is used for navigation.

The Sun's effect on the Earth's magnetic field

The Sun has a strong magnetic field, greatly exceeding that of any planet. It arises from convection in the solar core and is sufficiently irregular that it produces regions of lower than normal temperature on the Sun's surface, called sunspots. These affect the release of charged particles (electrons, protons, and alpha particles) from the Sun's atmosphere. The particles are not bound to each other, but form a plasma that spreads out at supersonic speed. The flow of electric charge is called the solar wind; it is accompanied by a magnetic field known as the interplanetary magnetic field. The solar emissions are variable, controlled by changes in the Sun's magnetic field. This reverses polarity every eleven years and causes a comparable frequency of sunspot activity, accompanied by increased emission of solar plasma. This regulates the intensity of the solar wind and affects its interaction with the magnetic fields of the Earth and other planets.

The magnetic field of a planet deflects the solar wind around it. This blocks the influx of solar radiation and prevents the atmosphere from being blown away, as may have happened to Mars and the Moon. Around the Earth (as well as the giant planets and Mercury) the region in which the planet's magnetic field is stronger than the interplanetary field is called the magnetosphere; its shape resembles the bow-wave and wake of a moving ship. In the case of the Earth it results when the supersonic solar wind, flowing at around 400 km/s, encounters the geomagnetic field 90,000 km above the Earth's surface. It compresses the field on the daytime side of the Earth, forming a bow shock, about 17 km thick, which deflects most of the solar wind around the planet. However, some of the plasma penetrates the barrier and forms a region called the magnetosheath; the boundary between the plasma and the magnetic field is called the magnetopause. The solar wind causes the magnetic field on the night-time side of the Earth to stretch out to form a magnetotail (Figure 30) that extends several millions of kilometres 'downwind' from the Earth. Similar features characterize the magnetic fields of other planets.

Although most of the solar wind passes by the Earth, some of the charged particles enter into the magnetic field and are trapped by it.

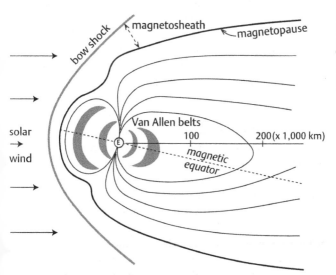

30. **Depiction of the Earth's magnetosphere (not to scale), illustrating the positions of the main features and the location of the Van Allen radiation belts. The multilayered ionosphere lies too close to the Earth's surface to be illustrated here.**

They form two large doughnut-shaped belts of radiation—the Van Allen belts—which girdle the Earth's magnetic axis, occupying the latitudes up to 20° on each side of the magnetic equator. The inner belt lies at distances of 1,000–3,000 km above the Earth, while the outer belt is at distances of 20,000–30,000 km. At one time it was thought that these radiation belts would present a hazard to spacecraft and their occupants but this does not appear to be the case. Very-short-wavelength radiation from the Sun is largely absorbed in the thin upper atmosphere, where X-rays, gamma rays, and ultraviolet radiation ionize air molecules.

The ionization spreads around the Earth, accumulating as concentric shells of charged particles to form the ionosphere. This consists of five layers in altitudes between about 60 km and 1,000 km above Earth's surface. The ionized layers reflect radio waves, which can travel around the world by multiple reflections between the ionosphere and the ground surface, enabling global radio communication. The flow of electrically charged ions in the ionosphere also causes magnetic fields that are detected on the Earth. They result in a small daily (or diurnal) variation in the magnetic field with an amplitude of 30–50 nT, large enough to require compensation during a magnetic survey. At times when the Sun's activity is high, sudden emissions of solar plasma result in large fluctuations in the magnetic field, called magnetic storms, which may necessitate interrupting a survey. On days when the Sun's activity is low (termed Sq days) the magnetic fields produced by the current systems in the ionosphere are used for electromagnetic investigation of the Earth's internal electrical conductivity. The Sq ionospheric fields with long periods in the range 10–1,000 s can induce electric currents (referred to as telluric currents) in conducting layers of the Earth. The telluric currents cause secondary magnetic fields that are measured on the Earth's surface and analysed to obtain information about the electrical conductivity of the crust, lithosphere, and upper mantle. The method is known as magnetotelluric sounding; it can be used for mineral and petroleum exploration, as well as in non-commercial investigations of the structure of the crust and upper mantle.

Satellite measurements of the geomagnetic field from the SWARM mission, together with ten years of observations by CHAMP, are so precise that it is possible to detect and use the secondary magnetic fields arising from oceanic tides. Seawater carries high concentrations of dissolved salts and is a good electrical conductor. The tides move the water through the Earth's magnetic field, which induce a weak electric current in the seawater. This in turn induces currents in the upper mantle, which produce magnetic fields recorded by the satellites. Interpretation of these signals enhances our knowledge of the electrical conductivity of the lithosphere and upper mantle below the oceans.

Magnetic fields of other planets

Mercury, the closest rocky planet to the Sun, has a global magnetic field that is close to being a dipole. However, the Messenger spacecraft found that Mercury's field is about 100 times weaker than the Earth's, and comparable in strength to the external magnetic fields due to the solar wind, which is strong because of Mercury's closeness to the Sun. As a result the planet's bow shock is very close to its surface.

Data from spacecraft that flew past Venus (e.g. Mariner 2 in 1962) or orbited the planet (e.g. Venus Express in 2006) have found that it does not have a significant magnetic field. This may be because it rotates so slowly: one day on Venus is equivalent to 243 Earth days. Rotation and the related Coriolis force, together with convection, are necessary factors for a self-sustaining dynamo, so possibly Venus does not rotate fast enough to generate a magnetic field. Mars and the Moon do not at present have global dipole magnetic fields, but orbiting satellites have detected magnetic fields related to magnetized areas of rock on their surfaces. This suggests that each may have possessed an internally generated magnetic field in the geological past. The Moon has a crust, mantle, and small core. However, the radius of the lunar core is only about 10 per cent that of the Moon, and the core is too small to support a dynamo. Satellite missions to Mars have measured intensely magnetized patches of crust but the planet does not have a global magnetic field that could be attributed to an internal dynamo.

The gaseous giant planets Saturn, Uranus, and Neptune have stronger central dipoles than the Earth, but these planets are so large that the magnetic fields on their surfaces are weaker than those on the Earth. In contrast, Jupiter has a huge magnetic field produced by a central dynamo in its core of liquid metallic hydrogen. Jupiter's dipole is 20,000 times stronger than the Earth's and occupies a larger share of the planet's interior. As a result Jupiter's non-dipole field components are strong relative to its dipole, which nevertheless predominates. The gaseous giant planet does not have a solid surface, but at the radial distance of its nominal surface the magnetic field is fourteen times stronger than on the Earth.

Magnetic properties of rocks

The Earth's magnetic field provides information about processes that are currently active deep within the Earth, and also helps scientists to understand the tectonic processes that have changed the planet in the geological past. The ability to analyse properties of the ancient magnetic field is based on the magnetic properties of rocks. A tiny fraction of the minerals in a rock—up to 1 per cent in some igneous and metamorphic rocks, but as low as 0.01 per cent in sedimentary rocks—possess magnetic properties similar to those of iron and are classified loosely as ferromagnetic. An important characteristic of these minerals is that they can be permanently magnetized in the geomagnetic field while they are being formed, acquiring a remanent magnetization. Equally important, the minerals can retain the remanent magnetization for very long periods of time.

Why are rocks able to act as magnetic recorders? The extraordinary stability of a magnetization in some rocks and sediments is due to the very fine grain size of the tiny proportion of magnetic minerals they contain, the most important of which are the iron oxides magnetite and hematite. An igneous rock forms from a hot fluid magma; as it cools, mineral grains crystallize and it solidifies. On further cooling magnetite and hematite grains in the rock pass through a critical temperature called the Curie point; below this temperature the grains behave like tiny magnets, with an important distinction. The grains are fixed in the solid rock, but the magnetizations inside the grains are able to align with and record the direction of the magnetic field in which it cools. Unless the rock is modified by some later process, its thermally acquired remanent magnetization is permanent. For example, the oceanic crust that forms at a spreading plate margin consists of basalt, an igneous rock, which becomes magnetized as it cools to ambient temperature in the geomagnetic field. Basalts sampled from oceanic crust that formed more than one hundred million years ago are still magnetized in the direction of the field in which they cooled.

Sedimentary rocks are also able to record the magnetic field in which they form. When a rock weathers and erodes, fine particles are transported by wind or water and deposited in a river bed, lake, or sea. Fine grains of magnetite in the waterlogged sediment can behave like miniature compass needles and align with the magnetic field during or shortly after deposition, thus acquiring a weak but stable magnetization that dates from the time of deposition. The remanent magnetization becomes locked into the sedimentary rock during the natural geochemical processes that caused it to harden. As in lavas, this magnetization also can survive for very long times. For example, the magnetizations of limestones from the Apennines and Venetian Alps still carry a faithful record of the geomagnetic field during the Late Jurassic, Cretaceous, and Palaeogene periods of geological time, an interval that lasted from 155 Myr to 25 Myr ago.

Palaeomagnetism

Investigations of secular variation recorded in the magnetizations of rocks and sediments indicate that, averaged over long geological time intervals, the dipole tilt and the NDF components have mean values close to zero. This means that the long-term magnetic field corresponds to a dipole at the centre of the Earth with its axis along the rotation axis. This is known as the geocentric axial dipole (GAD) hypothesis. It is the fundamental assumption of the discipline of palaeomagnetism, which has documented the motions of the continents relative to each other during hundreds of millions of years and has shown that the magnetic field has reversed polarity numerous times.

The palaeomagnetic method measures the direction of the remanent magnetization in rocks, specified by two angles: inclination and declination. The inclination is the angle between the direction of the GAD field and the horizontal surface of the Earth; the declination is the angle between the depositional magnetic north direction and geographic north. The inclination of the magnetization is the same as that of the field at which it was acquired and thus depends on the latitude (Figure 29). At the north magnetic pole the field points vertically downward and the inclination is 90 degrees; proceeding away from the pole the inclination gets progressively flatter until it is horizontal at the equator; in the southern magnetic hemisphere the field points upwards and has negative inclinations which steepen until the south magnetic pole where the direction is vertically upward. A simple equation describes this variation and allows the latitude of a place to be determined from the field inclination. Thus, if the inclination of the magnetization direction in a rock outcrop is determined, the latitude where the rock was formed can be calculated, and from this the distance to the place where the north magnetic pole was located at that time. The declination of the magnetization describes the direction from the sampling site to the pole. Thus, if one proceeds from the sampling location in the direction given by the declination for a distance calculated from the inclination, the location of the ancient geomagnetic north pole can be found. Due to plate tectonic motions since the rock was formed, the ancient pole position no longer lies on the present rotation axis and so it is known as the virtual geomagnetic pole (VGP).

Apparent polar wander and continental drift

Palaeomagnetic research on both igneous and sedimentary rocks of all ages has yielded average VGP locations for each age for each continent. It was observed early in these studies that the poles for a given continent appeared to move with time, defining a path of apparent polar wander (APW) that is unique to the continent. Moreover, each continent was found to have its own APW path, located approximately on the opposite side of the geographic pole from the continent. There is only a single geocentric axial dipole, which clearly could not have moved simultaneously along several different paths. Consequently, the different APW paths are evidence that the individual continents have moved relative to the rotation axis, a behaviour called continental drift. This slow motion of the continents was first proposed on the basis of different evidence early in the 20th century to explain the symmetry of shorelines and geological features across the Atlantic Ocean. The theory was contentious and set geophysicists against geologists because a suitable mechanism for moving the continents could not be found. Palaeomagnetic evidence for continental drift grew in the 1950s but the mechanism remained a mystery until the development of plate tectonic theory, which allowed the APW paths of individual continents to be recognized as traces of past plate motions.

The APW paths can be used to reconstruct earlier positions of the continents. For example, the palaeomagnetic poles of Europe and North America from Silurian times (about 440 Myr ago) to the middle Jurassic (about 170 Myr ago) define APW paths, which lie well apart in the present geographic framework (Figure 31A). The palaeomagnetic interpretation is that a single APW path formed while the respective plates were joined, and the present separation of the two paths is the result of later tectonic plate movements. As noted when discussing transform faults, the displacement of a tectonic plate on the Earth's spherical surface is equivalent geometrically to rotating the plate about an Euler pole of relative rotation.

To reconstruct the geological history of the Atlantic Ocean, a 38-degree clockwise rotation of the European plate about an Euler pole close to the present rotation axis brings the European APW path into a position where it overlaps the North American APW path (Figure 31B). This shows that the two continents were together from Late Ordovician (425 Ma) to Lower Jurassic (175 Ma), the length of time represented by the common segment of their APW paths. After the middle Jurassic, about 170 Myr ago, the Atlantic Ocean opened and the current phase of plate tectonic motions transported the European and North American continents to their present positions.

This type of analysis has been carried out for many of the continents. It has led to an understanding of the mobility of the Earth's lithospheric plates. In the distant past they collided to form supercontinents, which subsequently broke up and separated again. The agreement between the apparent polar wander curves (Figure 31B) documents the existence of a supercontinent, Euramerica. Similar results from the continents in the southern hemisphere show that they belonged to a southern hemisphere supercontinent, Gondwana, which subsequently broke up in a similar way. Euramerica and Gondwana were themselves portions of a single supercontinent called Pangea that encompassed all the

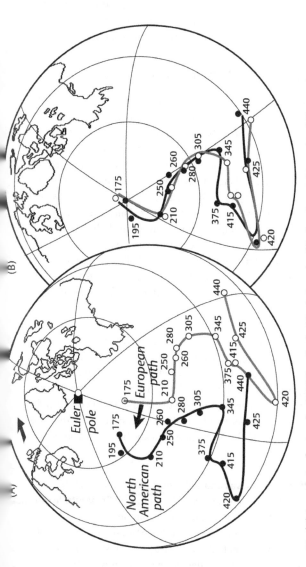

31. (A) Apparent polar wander (APW) paths for North America (black points and curve) and Europe (open points, grey curve) from Silurian to middle Jurassic age. (B) Overlap of the APW paths after rotating the European path by 38 degrees clockwise about a rotation pole (small square) close to the present geographic axis. The small numbers denote approximate ages in millions of years.

present-day continents and existed from about 300 Myr to 170 Myr ago. Pangea may have been preceded by other supercontinents but it is the most recent one and the best understood.

Geomagnetic polarity reversals

Early studies of the magnetizations preserved in lavas showed that half of the rock samples had normal magnetizations: that is, they had the same polarity as the present field. The remaining half had oppositely magnetized directions, and were defined as reversely magnetized. This evidence indicates that the Earth's magnetic field has changed polarity—from normal to reverse or the opposite—many times in the geological past. There have been roughly 300 reversals in the past 150 Myr alone, and an unknown number happened before then. The rate at which they occur has changed with time. In the interval 124–84 Myr ago the field did not reverse at all, forming a long interval of constant normal polarity. During the past 10 Myr there have been on average 4–5 reversals per Myr; the most recent full reversal happened 780,000 yr ago.

Exactly what causes the field to change polarity is not known, but it is a topic of current research into the geomagnetic dynamo. The power of supercomputers is required to construct models of how the magnetic field is created and how the process is affected by different factors, such as the growth of the solid inner core and the influence of the lower mantle D″ layer on convection in the fluid core. Palaeomagnetic results provide some boundary conditions for these computations.

Investigations in lavas and sediments have shown that the intensity of the dipole decreases by an order of magnitude during a polarity reversal. It does not vanish, but decreases to a low value and subsequently recovers as the new polarity is established. The process takes about 5,000–10,000 yr. At present, the intensity of the geomagnetic field is decreasing at roughly 5 per cent per century. This does not mean a reversal has started: the decrease in intensity could be a natural fluctuation.

During a polarity reversal, the direction of the field changes more rapidly than the intensity does, taking about 3,000–5,000 yr to reverse. On a human timescale the polarity transition is very slow, but on a geological timescale it is rapid. The polarity interval before and after a reversal is stable and persists much longer: it is called a polarity chron and may last from 50,000 to several million years. Sometimes the normal or reverse polarity chron is interrupted by shorter intervals of opposite polarity called subchrons. The pattern of magnetic reversals is irregular but distinctive for different times. This makes it possible for geologists to use the reversal pattern as a geological 'fingerprint' to date and correlate sedimentary rocks.

Igneous rocks acquire a thermal remanent magnetization during cooling from a molten state. They contain radioactive minerals that allow the rocks to be dated radiometrically, which makes it possible to construct a polarity-age timescale. Early investigations in lavas showed a characteristic sequence of normal and reverse polarity during the last few million years. The same polarity sequence was soon discovered in deep-sea sediments (Figure 32B), which have a depositional remanent magnetization acquired during deposition and sedimentation. Although the magnetizations were acquired by different mechanisms, the igneous and sedimentary polarity records agree, confirming that the polarity sequence is a feature of the geomagnetic field, rather than an artefact of the rock, which had been proposed by critics. The numerical ages of polarity reversals in both records are obtained by radiometric dating of the lavas. The dated polarity sequence is equivalent to a timescale that can be used to determine the ages of rocks and geological events. Numerous subsequent investigations have been made in cores drilled into the sedimentary layers on the ocean floor, as well as in outcrops of marine sedimentary rocks now exposed on land.

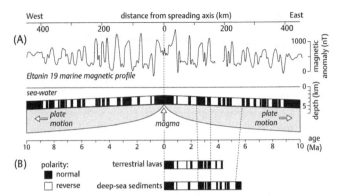

32. (A) Marine magnetic anomalies along a profile across the East Pacific Rise, and the interpreted record of geomagnetic polarity. (B) Geomagnetic polarity sequences from magnetic stratigraphy in terrestrial lavas and deep-sea sediments.

The determination of magnetic polarity in stratigraphically dated sediments is called magnetic stratigraphy. It has refined and extended the geomagnetic polarity timescale, so that this is now well known for most of the time since the breakup of Pangea around 170 Myr ago. This is possible because the polarity sequence found in lavas, sedimentary rocks, and oceanic sediments is also recorded in the magnetization of the oceanic crust formed since that time. The crustal magnetization disturbs the local geomagnetic field causing anomalies (Figure 32A) that can be measured by a magnetometer towed behind a ship in a marine magnetic survey.

Oceanic magnetic anomalies and plate tectonics

In the 1950s marine geophysicists found patterns of strong magnetic anomalies in the Pacific Ocean aligned as parallel lineations. Their origin was explained in 1963 by Frederick Vine and Drummond Matthews in Cambridge and independently by Lawrence Morley in Canada. The Vine–Matthews–Morley hypothesis relates the striped patterns of magnetic anomalies to the processes at oceanic ridges. The ridge is a divergent plate boundary, or spreading margin, where basaltic lavas are extruded that are strongly magnetized in comparison to other rocks. They acquire a thermal remanent magnetization in the direction of the ambient geomagnetic field in which they cool. Sea-floor spreading carries the magnetized oceanic crust away from the spreading axis. As a result, the oceanic crust is characterized by magnetized stripes, hundreds of kilometres long and 10–50 km wide and oriented parallel to the spreading ridge, that are magnetized alternately in opposite directions.

The magnetic field above normally magnetized crust has the same direction as the present field and reinforces it, making it locally stronger than average; the difference is a *positive* anomaly. Conversely, over reversely magnetized crust the field due to the magnetization is opposite to the present field and makes it weaker than average; in this case the difference is a *negative* anomaly. A magnetic survey across the magnetized stripes measures lineated magnetic anomalies with amplitudes of several hundred nanotesla (Figure 32A) corresponding to whether the underlying oceanic crust has normal or reverse polarity. The magnetic lineation patterns on opposite sides of a spreading ridge are symmetric. The anomaly pattern yields a record of the magnetic field reversals that happened during sea-floor spreading. The marine polarity sequence correlates well with polarity sequences found independently in lavas on land and in sediments from the ocean floor where they overlap (Figure 32B). These records from different sources mutually confirm and support each other. Moreover, because the ages of the lavas and sediments can be determined, the oceanic magnetic anomalies allow the ocean floor to be dated from the anomaly pattern, without needing to take rock samples.

By matching dated magnetic lineations on opposite sides of an oceanic ridge, geophysicists have been able to trace and time the motions of tectonic plates. For example, on a magnetic profile across a lineated set of anomalies on the Pacific–Antarctic ocean ridge (Figure 32A) sharp positive anomalies are evident 80 km east and west of the centre of the anomaly sequence. These were formed at the same time, about 2 Myr ago, when the opposite sides of the ridge were together at the spreading axis. Subsequent plate motion has separated them. If the plates adjoining the ridge are moved back together, so that the long stripes of magnetized crust corresponding to the same anomalies coincide, the positions of the plates 2 Myr ago can be reconstructed. Repeating this exercise for each dated pair of magnetic lineations reveals the history of relative motion of the plates and the continents that lie on them. The results are more detailed than the 'continental drift' reconstructions derived from matching palaeomagnetic APW paths. However, the method can only be applied to the present oceans, which are younger than 160 Myr. The palaeomagnetic method of matching APW paths provides reconstructions of ancient supercontinents for older epochs, albeit with less detail.

Chapter 8
Afterthoughts

Geophysics has made many important advances towards
understanding the behaviour and properties of planet Earth.
Geophysical research continues to make discoveries and
developments that benefit society. Many problems presented by
the Earth—such as the prediction of both the time and location
of an earthquake—have been researched by geophysicists for
decades but remain unsolved because of the complex nature
of their causes.

Many sophisticated geophysical techniques were developed
originally for commercial purposes by the oil and mineral
industries but have subsequently found their way into general
use. On a small scale, the field of environmental geophysics uses
seismic, magnetic, gravimetric, and electromagnetic methods to
investigate the shallow underground (the top 100 m or so) and
resolve issues involving the environment; typically these might
be investigations of geological hazards, the groundwater, or
archaeological sites. At the other extreme of scale and expense,
numerous satellite missions have been carried out with
geophysical goals, and have delivered abundant amounts of new
information about the planet, complementing and augmenting
existing knowledge, especially about the gravitational and
magnetic fields.

The advances in methodology are illustrated by the observation and measurement of rates of plate motion. These rates were once based solely upon the analysis of magnetic anomalies at spreading ridges, but are now supplemented by space-geodetic observations made from satellites. Long-term measurements at VLBI stations on different continents have recorded the changing separations of some continents that result from plate tectonics. For example, the distance between VLBI stations in North America and Europe is increasing steadily at a mean rate of 17 mm/yr (about half as fast as fingernails grow). The best-known space geodetic technique, GPS, when deployed in global networks of receivers, can observe plate motions directly, even detecting spreading rates at incipient plate boundaries—such as the East African rift zone—which have not yet formed a classical spreading ridge.

As in other sciences, new hypotheses are not always accepted initially in geophysics. This is often because of political and human factors. Ptolemy's geocentric model of the solar system used artificial constructs such as 'epicycles' and 'equants' to account for anomalous planetary motions; encouraged by religious dogma, the model was held to be true for more than 1,500 years. It was only replaced by the heliocentric model of Copernicus after irrefutable evidence from the planetary observations of Kepler and Galileo became widely accepted. In a like manner, many earth scientists only accepted the theory of plate tectonics when the results of deep-sea drilling in the Atlantic Ocean provided hard evidence that the age of the oceanic crust increased with distance from the mid-oceanic ridge. Similar academic scepticism greeted the hypothesis of Luis and Walter Alvarez that the extinctions of many forms of life at the end of the Cretaceous Period (66 Myr ago) resulted from the impact of an extraterrestrial object, an asteroid or comet. The impact theory finally became generally accepted when geophysicists located the crater produced by the impact at Chicxulub, off the coast of the Yucatan peninsula in Mexico. Similarly, the existence of mantle plumes that feed hotspots has been a controversial topic. Most geodynamicists favour the plume concept, which is compatible with observed geoid and topography anomalies over the hotspot locations, but some seismologists maintain that the available data do not fit plume theory. However, analysis of hotspots indicated that there are three different types; some are tied by plumes to the D″ layer at the core–mantle boundary, some are related to the lower mantle superplumes under Africa and the Pacific, and others are linked with a shallower origin. Based on a global shear wave velocity model of the mantle, recent seismic analysis has shown vertically continuous low-velocity columns under some—but not all—hotspots. Although broader than the expected thin conduits, they lend seismic support to the plume theory for these hotspots.

The Earth is a bountiful laboratory for scientific investigations. Geophysics has always been a key player in developing an understanding of how the Earth works. It has an indispensable role in the search for the material resources that are needed for the smooth functioning of modern society. Geophysics helps to locate the fossil fuels needed to supply our energy needs, and the rare earth metals that are essential components of everyday devices, such as mobile phones and computers. On the other hand, geophysics is an important factor in protecting society against natural hazards by monitoring earthquakes, volcanoes, and unstable slopes, and by warning against possible tsunamis and the negative effects of solar flares on electrical networks.

Due to the complexity of many natural topics, geophysical investigations often do not yield data that fully comply with a theory. Earth scientists cannot control the natural events that occur, but can only observe and try to understand them. The equipment for attaining these goals improves continually, in a remarkably rapid and encompassing manner. Land-based methods are now augmented in several fields by remarkable observations from space. In some cases multiple sources of evidence are brought together to clarify an important issue. For example, palaeomagnetism documents plate motion across a stationary hotspot; satellite and marine observations of gravity, together with oceanic bathymetry, record the effects of uplift; while seismology and geochemistry contribute to solving the geodynamic basis of the observations. The processes at work in the Earth are complex and their causes are well hidden, but this provides each discipline of geophysics with interesting challenges to engage present and future generations of scientists.

Further reading

Most of the following books are written at an introductory level and are suitable for a lay reader. They include general books on the earth sciences, in which geophysical topics are handled within a larger framework of information about the Earth. Some books describe specific geophysical topics in more detail than is possible in the compressed format of this book. Three are textbooks, perhaps more suited to students who intend to continue a career in earth sciences, but which contain chapters that provide useful reference material and more in-depth knowledge than contained in this book.

Introductory books on the earth sciences

G. C. Brown, C. J. Hawkesworth, and R. C. L. Wilson (eds) (1992), *Understanding the Earth*, Cambridge University Press, Cambridge, 551 pp.

F. Press, R. Siever, J. Grotzinger, and T. Jordan (2003), *Understanding Earth*, 4th edn, W. H. Freeman, San Francisco, 568 pp.

M. Redfern (2003), *The Earth: A Very Short Introduction*, Oxford University Press, Oxford, 141 pp.

Specific topics in geophysics

B. A. Bolt (2003), *Earthquakes*, 5th edn, W. H. Freeman and Co., New York, 320 pp.

R. F. Butler (1992), *Paleomagnetism: Magnetic Domains to Geologic Terranes*, Blackwell Scientific, Boston, 319 pp.

G. F. Davies (1999), *Dynamic Earth: Plates, Plumes and Mantle Convection*, Cambridge University Press, Cambridge, 470 pp.

P. Kearey, K. A. Klepeis, and F. J. Vine (2013), *Global Tectonics*, 3rd edn, John Wiley and Sons, New York, 496 pp.

P. Molnar (2015), *Plate Tectonics: A Very Short Introduction*, Oxford University Press, Oxford, 136 pp.

Geophysical textbooks for reference or more in-depth reading

C. M. R. Fowler (2004), *The Solid Earth: An Introduction to Global Geophysics*, 2nd edn, Cambridge University Press, Cambridge, 500 pp.

W. Lowrie (2007), *Fundamentals of Geophysics*, 2nd edn, Cambridge University Press, Cambridge, 381 pp.

A. E. Mussett and M. A. Khan (2000), *Looking into the Earth: An Introduction to Geological Geophysics*, Cambridge University Press, Cambridge, 492 pp.

Geophysics

Index

Index

Geophysics

Geophysics

DESERTS
A Very Short Introduction
Nick Middleton

Deserts make up a third of the planet's land surface, but if you picture a desert, what comes to mind? A wasteland? A drought? A place devoid of all life forms? Deserts are remarkable places. Typified by drought and extremes of temperature, they can be harsh and hostile; but many deserts are also spectacularly beautiful, and on occasion teem with life. Nick Middleton explores how each desert is unique: through fantastic life forms, extraordinary scenery, and ingenious human adaptations. He demonstrates a desert's immense natural beauty, its rich biodiversity, and uncovers a long history of successful human occupation. This *Very Short Introduction* tells you everything you ever wanted to know about these extraordinary places and captures their importance in the working of our planet.

www.oup.com/vsi

LANDSCAPES AND GEOMORPHOLOGY
A Very Short Introduction
Andrew Goudie & Heather Viles

Landscapes are all around us, but most of us know very little about how they have developed, what goes on in them, and how they react to changing climates, tectonics and human activities. Examining what landscape is, and how we use a range of ideas and techniques to study it, Andrew Goudie and Heather Viles demonstrate how geomorphologists have built on classic methods pioneered by some great 19th century scientists to examine our Earth. Using examples from around the world, including New Zealand, the Tibetan Plateau, and the deserts of the Middle East, they examine some of the key controls on landscape today such as tectonics and climate, as well as humans and the living world.

www.oup.com/vsi